GOVERNING WATER IN INDIA

Governing Water in India

Inequality, Reform, and the State

LEELA FERNANDES

UNIVERSITY OF WASHINGTON PRESS
Seattle

The open access edition of *Governing Water in India* was made possible in part by a grant from the University of Washington Libraries and the Libraries Excellence Fund. Additional support was provided by generous gifts from Vandana Nandan and Vikram Nagaraj.

Copyright © 2022 by the University of Washington Press

Portions of chapter 3 were previously published in Leela Fernandes, "Inter-state Water Disputes in South India," *Oxford Research Encyclopedia in Asian History*, April 26, 2018, https://doi.org/10.1093/acrefore/9780190277727.013.191.

Composed in Minion, typeface designed by Robert Slimbach

The digital edition of this book may be downloaded and shared under a Creative Commons Attribution Non-Commercial No Derivatives 4.0 international license (CC-BY-NC-ND 4.0). For information about this license, see https://creativecommons.org/licenses/by-nc-nd/4.0. This license applies only to content created by the author, not to separately copyrighted material. To use this book, or parts of this book, in any way not covered by the license, please contact University of Washington Press.

Photos by author. Maps and figures by Ben Pease, Pease Press Maps.

UNIVERSITY OF WASHINGTON PRESS
uwapress.uw.edu

LIBRARY OF CONGRESS CATALOGING-IN-PUBLICATION DATA
Names: Fernandes, Leela, author.
Title: Governing water in India : inequality, reform, and the state / Leela Fernandes, University of Washington Press.
Description: Seattle : University of Washington Press, [2022] | Includes bibliographical references and index.
Identifiers: LCCN 2021050716 (print) | LCCN 2021050717 (ebook) | ISBN 9780295750422 (hardcover) | ISBN 9780295750439 (paperback) | ISBN 9780295750446 (ebook)
Subjects: LCSH: Water-supply—Government policy—India. | Economics—India. | Equality—India.
Classification: LCC HD1698.14 F46 2022 (print) | LCC HD1698.14 (ebook) | DDC 333.9100954—dc23/eng/20211227
LC record available at https://lccn.loc.gov/2021050716
LC ebook record available at https://lccn.loc.gov/2021050717

♾ This paper meets the requirements of ANSI/NISO Z39.48-1992 (Permanence of Paper).

CONTENTS

List of Illustrations   vii
Preface and Acknowledgments   ix
List of Abbreviations   xiii

INTRODUCTION   1
CHAPTER 1.   The Historical Formation of India's Water Bureaucracy   29
CHAPTER 2.   The Regulatory Water State in Postliberalization India   69
CHAPTER 3.   The Political Economy of Federalism and the Politics of Interstate Water Negotiations   96
CHAPTER 4.   Regulatory Extraction, Inequality, and the Water Bureaucracy in Chennai   142
CHAPTER 5.   State, Class, and the Agency of Bureaucrats   192
CONCLUSION   225

Notes   239
Works Cited   251
Index   273

ILLUSTRATIONS

Figures

4.1 Chennai's Water Bureaucracy   148
4.2 Water Tanker in Chennai   173
4.3 Tamil Nadu Housing Development Water Source   188

Maps

1.1 The Madras Presidency in Colonial India   34
3.1 Southern Indian States   100
3.2 The Cauvery River   108
3.3 The Telugu Ganga/Krishna Water Supply Project   126
3.4 The Kerala Mullaperiyar Dam   130
4.1 Chennai's Water Supply System   158

Tables

1.1. PWD engineering staff in colonial India   44
1.2. Tamil Nadu PWD irrigation budget, 1967–1973   61
4.1. Urbanization in Tamil Nadu   149
4.2. Sectoral water demand, Tamil Nadu   167

PREFACE AND ACKNOWLEDGMENTS

The twenty-first century has been marked by an unprecedented crisis of governance. In the twentieth century, proponents of programs and theories of modernization have posited a relationship between democracy and development as the ideal path forward for postcolonial nation-states. This postulation was reformulated in the 1980s into what became known as the "Washington Consensus." The Washington Consensus posited policies of economic reform broadly geared toward privatization, liberalization, and a reduced role of the state. The post-Washington consensus shifted to include and foreground institutional reforms that were held up as the means for both economic growth and efficient and accountable governance. Yet nation-states in both the Global North and the Global South have witnessed crises as exclusionary governments often linked to right-wing populist movements have sought to hollow out democratic institutions and curtail access to democratic rights and political participation. Such crises have taken root at the moment when global challenges of climate change and local and global inequalities make effective and accountable governance one of the most pressing issues of our time. There has been no better illustration of this than the global health pandemic of COVID-19 that at various stages has acutely highlighted the stakes of ineffective governance in countries as varied as India, Brazil, and the United States.

This book grapples with the question of governance through a study of water in contemporary India. Understanding the constraints and possibilities of effective and accountable governance compels us to wrestle with complex, historically situated local, national, and global configurations in particular places. Such analyses are not glamorous, and they do not draw in the reader with the spectacle of the suffering of marginalized communities or the comfort of abstract, modular social scientific policy responses. Rather, *Governing Water in India* invites the reader to grapple with the messy, imbricated processes that create, perpetuate, and worsen the devastating effects of water scarcity and the intensifying cycles of phenomena such as droughts and floods. Most significantly, the book seeks to disrupt the decoupling of environmental phenomena from policies of economic and institutional reform. Such environmental challenges are deeply connected to successive dominant models of economic development and, in the twenty-first century, to growth-oriented policies of economic reform. Responses to global challenges of climate change and inequality require both systemic analyses that address the impact of such policies and located studies of particular places and contexts. Responses to such challenges will ultimately require a deep understanding of the institutional contexts that will determine whether the implementation of policy prescriptions is successful or ineffective.

In this endeavor, *Governing Water in India* presents in-depth analyses of understudied institutions that make up local water bureaucracies. I am grateful to the Tamil Nadu Public Works Department and its various water institutions for providing me with access to its organizational archives. In one case, this archive consisted of a room full of documents, which included original colonial documents that almost crumbled as I turned the pages to governmental and international organization reports and assessments. Waiting for the PWD employee to break open some of the locks was a telling experience during my field research. The documents had not been accessed, and the employee did not have up-to-date keys for all the cabinets, so he was somewhat miserable when in response to his query I told him I wanted to read everything. The closed-off treasure trove of this hidden archive underlined the paradox of how much is written about bureaucracy, governance, and corruption and how little interaction there is between academics and such organizations that are actually designated with the implementation of policies. With this recognition in mind, this book is written in a style to make it accessible to a range of academic and public audiences in India and in comparative contexts. I have minimized rhetorical embellishments that sometimes nourish US academic desires for stylistic flourish in favor of accessibility and the commitment to foreground details that may matter to individuals grappling with the realities of governing water even if they do not seem to be of import to faraway academic audiences.

Fieldwork for this research took place in Chennai between 2016 and 2018. I am grateful to the Centre for Water Resources at Anna University for providing me with access to its library resources. Thanks go to numerous people for their intellectual engagement and logistical support of the field research: Dr. N. K. Ambujam, Dr. K. Ilamparuthi, Prof. S. Janakarajan, and Prof. R. Mahalingham. I am also grateful to members of several organizations whom I keep anonymous in accordance with research ethics, including those from Metrowater, PWD, TNUIFSL, Tamil Nadu's Smart Cities Mission, Tamil Nadu Employment and Exchange Offices, and the Tamil Nadu Slum Clearance Board. Funding for this research was provided by the University of Michigan. I am also especially grateful to Saisha Nanduri for her excellent research assistance and all of the work she did in putting together the book's bibliography.

The arguments in this book were improved by feedback that I received from numerous venues. I am particularly grateful for two rounds of feedback from the Centre for Political Studies at Jawaharlal Nehru University—first at their center's seminar and later when I presented comprehensive arguments as the their annual Nirman Foundation speaker. My special thanks to Asha Sarangi for organizing these events. The book benefited from feedback from discussions at Johns Hopkins University, the Michigan Society of Fellows, the University of Pittsburgh, and Shiv Nadar University. The book benefited from supportive editorial comments from my editor Lorri Hagman at the University of Washington Press. I am also especially grateful to the University of Washington Libraries for their generous support, which has allowed this book to be accessible as an open source book. My thanks also go to my colleague Anand Yang for suggesting that I submit the manuscript to UW Press. I am grateful to numerous colleagues for their professional support, including Amrita Basu, Aimee Germain, Ruth Wilson Gilmore, and Nancy Naples. This book could not have come to fruition without the tremendous and long-standing personal support from Nikol Alexander-Floyd, Jane Junn, and Asha Sarangi. They have been constant models of the courage and persistence needed to develop ethical governance within our own institutions in the academy. Finally, this book relied on my precious community outside of the academy, whose members both nourished my spirit through the earthly joys of clay and continually reminded me that water matters because it ultimately transcends human control and willpower and teaches us humility.

ABBREVIATIONS

AAP	Aam Aadmi Party
ADB	Asian Development Bank
AIADMK	All India Anna Dravida Munnetra Kazhagam
BJP	Bharatiya Janata Party
CGWB	Central Ground Water Board
CMC	Chennai Metropolitan Corporation
CMDA	Chennai Metropolitan Development Authority
CMWSSB	Chennai Metropolitan Water Supply and Sewerage Board (Metrowater)
CWC	Central Water Commission
DMK	Dravida Munnetra Kazhagam
GoTN	Government of Tamil Nadu
IAS	Indian Administrative Service
MCM	million cubic meters
mha	million hectares
MLA	Member of the Legislative Assembly
MMC	Madras Metropolitan Corporation
MMDA	Madras Metropolitan Development Authority
OBC	Other Backward Class
PPP	public-private partnership
PWD	Public Works Department
SC	Scheduled Caste
TMC	thousand million cubic feet
TNSCB	Tamil Nadu Slum Clearance Board
TNUDF	Tamil Nadu Urban Development Fund
TNUIFSL	Tamil Nadu Urban Infrastructure Financial Services Limited
ULB	urban local body
UNDP	United Nations Development Programme
WRCP	Water Resources Consolidation Project
WRO	Water Resources Organisation
WUA	Water Users' Association

GOVERNING WATER IN INDIA

Introduction

We should work for conserving rainwater. Panchayats should spend NREGA [National Rural Employment Guarantee Act] fund on water conservation during April. . . . Every drop of rain should be conserved. This would not only save villages from a water shortage. This would also help in agriculture.
—Prime Minister Narendra Modi

Furthermore, India ranks 133rd out of 180 nations for its water availability and 120th out of 122 nations for its water quality. It has been evaluated that 80 percent of India's surface is polluted which results in India losing US$ 6 billion every year due to water-related diseases. Challenges faced by the Indian water sector are due to increasing water consumption and wastage in urban areas, water-borne diseases, industrial growth, political and regulatory disputes, water cycle imbalances, increasing irrigation and agricultural demand, lack of technology, etc. According to estimates, India's water sector requires investment worth US$ 13 billion.
—Government of India Smart Cities Mission

GOVERNANCE OVER WATER IN INDIA HAS BECOME A FORMIDABLE endeavor for the Indian state. Water is essential for human life and therefore a fundamental need that the government must ensure is fulfilled for its citizens. Water is also a resource that is in demand from competing sectors of the economy. The state's administration of water resources and infrastructure involves governmental action across all levels of India's federal structure and consequently illuminates every facet of the Indian state. The complexity of the state's approach to the governance of water has been further deepened in the postliberalization period, which began in the 1990s, by the effects of policies of reform, accelerated economic growth, and unplanned urbanization. Older approaches of the developmental state have intersected with and been reconfigured in the postliberalization period in ways that have deepened the strains on water resources and produced new challenges for governance at the local, regional, and national levels.

Consider two snapshots of these challenges that are captured by the pronouncements of Prime Minister Narendra Modi and the Smart Cities Mission of the Indian government. In a visit to promote developmental activities in Madhya Pradesh, Modi addressed water shortages that villages were experiencing. Speaking to a group of villagers and members of tribal communities in Madhya Pradesh, he called on the village *panchayats* (local elected governing bodies) to redeploy funds from the state's employment assistance program (MGNREGA) to address water conservation. Modi's rhetoric is emblematic of the weighty language of India's conventional developmental state. The use of funds from the governmental program to meet dire rural water needs is a classic instance of the state's approach of using such assistance to meet a basic need such as the provision of water.

The prime minister's rhetoric on the need for intensive water conservation also combined these deep-rooted developmental languages with newer rhetoric associated with the postliberalization period. If his exhortation to villagers and tribal communities that "every drop of rain should be conserved" invokes the duty and sacrifice of citizens that the developmental state has echoed since the 1950s, it also now reflects the stress on decentralized local governance and self-help approaches of the postliberalization period.[1]

This shift from the top-down developmental state languages and approaches to governance over water is also embodied in other state endeavors in the postliberalization period. Consider, for instance, one of the key governmental schemes of the Modi government, the Smart Cities Mission, which was designed to develop and modernize India's cities. The Mission presents a stark portrait of the strains on governance over water "due to increasing water consumption and wastage in urban areas, water-borne diseases, industrial growth, political and regulatory disputes, water cycle imbalances, increasing irrigation and agricultural demand, lack of technology, etc." The delineation of the challenges of the deepening multiple demands on water resources by the government's Smart Cities Mission and the economistic conception of water resources as a sector of the economy in need of $13 billion reflects both the discursive shifts and policy challenges of the postliberalization period. Higher economic growth has of course placed greater pressure on water resources. More significantly, the organizational architecture of the Smart Cities Mission itself embodies both the institutional and the policy changes of the postliberalization state that has increasingly been shaped by a city-centered model of urbanized growth.[2]

The gravity and scale of such challenges—ranging from local water shortages and conservation practices in villages to the systemic needs of city-centered growth—illustrate the ways in which the governance of water represents a critical site for an understanding of the Indian state. The multifarious nature of water provides a distinctive analytical terrain in which we can disentangle the various facets of the postliberalization state. The politics and political economy of water are shaped by an intricate configuration of historical legacies stemming back to the colonial period, state polices of the twentieth-century developmental state, and long and varied histories of political action, negotiation, and conflicts both within and between civil society and the state. The subject of water has been shaped in fundamental ways by the dynamics of Indian federalism as it has fallen under the purview of local, state, and central governmental authority and has been a critical subject for judicial intervention by India's Supreme Court. It is within this variegated political, economic, and institutional field that policies of institutional reform have targeted India's frameworks of governance over water resources.

Reforms of the institutional architecture of water governance have rested on the two foundational principles that have become dominant national principles of reform—decentralization and privatization. These principles have been encoded in new national water policies of the central government, as well as in policies of restructuring that have been taken on by central and state governments across India. They have also converged with the norms of global institutions such as the World Bank and the Asian Development Bank, which have been central players in the water sector in comparative contexts (Bakker 2010; Morgan 2011). These principles and policies of reform have had a significant impact on the governance of water at various levels of state authority within India's federal structure (Kumar 2009). However, processes of reform and restructuring have also belied the conventional narrative that has foregrounded both positive and negative effects of privatization and the role of private capital in the water sector. Instead, the case of water sector reforms reveals a set of contradictory processes in which policies of reform have both reoriented as well as continued and intensified longstanding practices and modes of state power.

In contrast to the rhetoric on decentralization and privatization, processes of reform have intensified state centralization. Such processes both build on and reconfigure the historical weight of the institutional legacies of the colonial and developmental state. Institutional reforms designed to scale back the role of the state through processes of decentralization and the participation of private sector actors in fact produce a redistribution of centralized institutional power. This rethinking of processes of state centralization deepens our understanding of two key debates on the postliberalization state. First, processes of centralization cannot be understood purely as a product of a monolithic and recalcitrant bureaucracy. While bureaucrats often, unsurprisingly, attempt to retain their power and authority over resources, those who do attempt to perform effective regulatory functions are often constrained by structural conditions of the political economy of their institutions. More significantly, processes of state centralization are intrinsic to processes of economic reform that have been unfolding in late twentieth- and twenty-first-century India. The dynamics of institutional reform and processes of centralization are significant factors in the reproduction of socioeconomic inequality. A focus on such institutional dynamics moves us away from conventional accounts of neoliberalism that often identify privatization as the sole or determinant factor shaping inequality.

Institutional Reforms, Bureaucratic Agency, and Inequality

The governance of water encompasses a vast set of issues that include meeting the basic demands of citizenship by providing safe and accessible drinking water, managing agricultural and industrial needs for water resources, maintaining and expanding water infrastructure, and responding to crises that stem from droughts and floods (Ballabh 2008). In contemporary India, competing demands over water and decades of inadequate governance have transformed water into a site for significant political conflicts within localities, between state governments, among users (for instance rural and urban communities), and between the state and citizens (see Asthana and Shukla 2014; Bandyopadhyay 2016; Baviskar 2005; Cullet 2009; Iyer 2015; Mohan, Routray, and Sashikumar 2010; Shah and Vijayshankar 2016). The challenges for water governance have only intensified as unplanned urbanization and industrial investment, along with the impacts of climate change, produce new strains on water resources.

The governance of water in postliberalization India rests at the conjuncture of two processes. On the one hand, national and local regulatory reforms (in keeping with dominant global norms) have identified privatization and decentralization as the central pillars for the effective management of water resources.[3] Institutional and economic reforms produce new forms of centralization that are an integral part of the policies being implemented. An understanding of how this process works requires a broadening of the concept of centralization that moves beyond spatialized conceptions of a center-state framework of governance. Spatialized conceptions tend to associate centralization with an upward devolution of power from the local, state, and central governments. However, in practice, the dynamics of centralization have more to do with the concentration of power, authority, and control over resources. For instance, effective regulatory processes of the central government are necessary for successful processes of decentralization. Regulatory frameworks from the central government are necessary for effective governance and are distinct from centralized modes of state authority. Meanwhile, the concentration of authority over resources by state governments can intensify centralized state authority at the local level by strengthening vertical mechanisms of state power over local communities. Furthermore, such local governance structures can also reproduce the hierarchical relationship between social groups (often based on caste, gender, and religion) within civil society.

Recent scholarship has illustrated the ways in which state governments have sought to limit the autonomy of local governments, thus curtailing decentralization at the local level (Lobo, Sahu, and Shah 2014; Sangita 2014; Sharma and Swenden 2017; Vaddiraju 2014). State governments have increasingly sought to curtail the decentralization of power to local governments and have often either blocked or reoriented reforms in ways that have reinforced the concentration of authority within the institutions of state government. This concentration of state authority is organized around the growing power of state governments and the realm of city-centered modes of urban governance in particular. At one level, state governments are able to exercise political and institutional control over local governments through the control of finances and the appointment of officials (Sangita 2014, 88). At another level, the postliberalization economic model has produced new forms of centralized power within states. Political scientist Anil Kumar Vaddiraju has noted that state governments consolidate their own authority within state capitals and do not decentralize the control over decision-making and resources to the third level of local governance (2014, 101). While the ideational model of economic reforms proposes a reduced role of the state both by restraining state intervention in the economy and by promoting political and economic decentralization, such processes in fact produce new forms of the centralization of power that occurs at various spatial scales. For example, an increase of centralization occurred through the increasing power of chief ministers (Manor 2016). In other words, centralization is not simply about the concentration of power at the largest scale of government (the central government) but about processes of centralization that are reworked across multiple scales of state authority.

State capitals are an example of a central locus of new modes of urban governance in which state authority is located and centralized in the postliberalization period (Sassen 2001). Such sites of urban governance are instances of the new state spaces that are a central technology of state power in the context of contemporary processes of globalization (Brenner 2004). While this book is about urban bureaucratic governance, its implications are of relevance to scholars interested in local programs of decentralization. Its analysis of urban governance is not framed as a territorialized image of a city bureaucracy; rather, it is a systemic analysis of the redistribution and reconsolidation of state power. Its account of city bureaucracy is as much a story about rural India as about the challenges of urbanization and urban governance. Programs and policies of decentralization that can range from formal governmental programs of centralization to grassroots community-based models of governance take root in conjunction with these centralized processes. The languages and programs of decentralization in effect mask this centralization of state power.

Reforms of the governance of water produce a redistribution of state power that is shaped by the ascendancy of a city-based model of development. Water reforms provide a rich case for a systematic analysis of these processes of centralization. Some reforms encode some city-oriented state agencies with new forms of centralized authority, while policies of decentralization target small towns and rural areas in ways that both reflect the political and economic weaknesses of and intensify the control of state governmental authorities of these areas. In this process, regulatory reform is transformed into a process of regulatory extraction that encodes relationships of power both within and between urban and rural communities. While dominant rural groups continue to hold both socioeconomic power and political power in terms of electoral calculations of political parties, the urban-oriented effects of liberalization intensify the divisions and inequalities between urban and rural communities. In this transition, the state does not abandon but restructures its welfarist framework. The regulatory state thus produces a redistributive shift that accentuates long-standing socioeconomic inequalities while deepening new divides between larger urban areas and smaller rural and urban towns. Research on comparative urban localities in India has revealed systematic patterns of urban appropriation of water that was primarily used for agricultural purposes (Celio, Scott, and Giordano 2010; Punjabi and Johnson 2018). Varying institutional arrangements shape such patterns of appropriation. This book provides an in-depth analysis of *how* such institutional patterns are being shaped by reforms of the governance of water.

Such patterns of institutional change have been shaped by the national dynamics of India's centralized model of federalism (Sharma and Swenden 2017). This has meant that domestic political processes and institutions have shaped reforms in ways that refashion or compromise idealized norms of the regulatory state (Dubash and Morgan 2013; Jenkins 2004; Manor 2004; Mooij 2005). Analyses of the limits of regulatory reform in India have focused on the ways in which domestic political processes, institutional resistances, and corruption serve as roadblocks to institutional reform.[4] While building on the insights of such scholarship, this book also questions the presumption that regulatory reform would lead to decentralization if not for the constraints of preexisting political, socioeconomic, and institutional interests and resistances. Academic and public narratives on bureaucratic corruption and politicized institutional dysfunction often inadvertently produce an argument of Indian exceptionalism. In such a conception, it is the specificities of the Indian political and institutional context that hinder an idealized model of reform from being effectively implemented. While domestic political factors and bureaucratic resistance and corruption remain an integral part of an explanation of why regulatory reform falters in India (Bussell 2012), they do not fully account for the ways in which the centralizing policies inherent in the policies of institutional and economic reform produce regulatory failures. In order to move away from such assumptions of Indian exceptionalism, a more complex analysis of bureaucratic agency is needed. Analysis of the ways in which bureaucrats are themselves enmeshed in such broader and differentiated institutional political and socioeconomic fields (Bourdieu 1994) opens up the conceptual space for an understanding of the structural constraints on bureaucrats and the ways in which bureaucrats seek to navigate them. Bureaucrats are complex actors who have often sought to effectively carry out their regulatory responsibilities. However, hierarchies and structures of power within the broader field of state institutions curtail the potential for effective bureaucratic action.

By focusing on the fine-grained texture of how state water institutions work and how institutional reforms produce and reproduce new forms of extractive relationships, this book contributes to an understanding of how inequalities are produced in the postliberalization period. A sizable body of scholarship has illustrated the ways in which access to water resources is shaped by long-standing intersecting inequalities of caste, class, and gender (Ballabh 2008; Mehta 2013). Anthropological scholarship in particular has yielded rich understandings of how state power and inequality intersect in metropolitan cities such as Mumbai, Bengaluru, Calcutta, and Chennai (Anand 2017; Björkman 2015; Coelho 2017; Dasgupta 2015). Such findings are borne out in my research on institutional reform. What such research points to is a need to understand the relationship between institutional reforms, domestic political processes, and the reproduction of socioeconomic inequalities such as caste, class, and gender. What is at stake here is not the conclusion that water resources are not distributed in an equitable manner or that inequality is produced or reproduced through water distribution systems. Rather, the point at hand is to understand *how* institutional reforms are shaped by such inequalities and *how* inequalities are produced by institutional practices. The reworking of centralized state power through regulatory reforms provides the new institutional nodes for such processes of extraction and the reproduction of the enduring forms of inequality that have characterized contemporary India.

Concerns about water scarcity and phenomena such as the overextraction of groundwater in India have produced major governmental reports and proposed regulatory frameworks at the national level. Mihir Shah's (2016) report on the restructuring of two key institutions, the Central Water Commission (CWC) and the Central Ground Water Board (CGWB), for instance, presents a broad and intensive assessment of the challenges of water governance and a need for a paradigm shift in the state's approach to water resources. The report provides a comprehensive overview of the need for an institutional framework that can address varied and complex challenges, such as the need to increase irrigation efficiency, address pollution, deepen participatory decentralization, rejuvenate rivers, develop sufficient sewage treatment facilities for cities, broaden the disciplinary training of water bureaucrats, and address the growing effects of climate change.

In the midst of the report's outline of this vast set of challenges and institutional responses, three points merit foregrounding. First, while the report is tasked with addressing the reform of the two centralized institutions, the CWC and CGWB, the role of state governments in meeting governance challenges remains a cornerstone for the implementation of any proposed reforms. The report points to the need to "incentivise and facilitate" state governments' reforms (43). Second, the report argues that a paradigm shift in the governance of water requires a break from long-standing bureaucratic frameworks that approach water governance through "techno-centric supply-side interventions implemented top-down by fragmented bureaucracies, involving mostly technology, engineering and public investment in water infrastructure" (21). Finally, while substantial parts of the report are devoted to irrigation, it notes that in the context of expanding processes of urbanization and industrialization, it has become increasingly critical for national institutions such as the CWC to adopt "a holistic view of the often competing and conflicting demands of urban and rural areas, as also agriculture and industry" (16). A focused analysis of the role of state governmental is thus critical for an understanding of the governance of water in twentieth-century India.

Tamil Nadu provides a rich case for an analysis of institutional reforms. The state embodies a key site for an analysis of the historical emergence of both India's water bureaucracy and the political and economic structures of the colonial and developmental state. Such historical processes have shaped the regional political economy of water in southern India in ways that are illustrative of the strains on the federal governance of water. The historical significance of the state allows for an analysis of the historical continuities that shape and constrain postliberalization reforms. Among Indian states, Tamil Nadu has been a leader in implementing global and national reforms centered on the principles of decentralization and privatization (Harriss and Wyatt 2019). However, the state's reform agendas have been shaped by a strong water bureaucracy that has incorporated processes of decentralization within its centralized bureaucracy. The Tamil Nadu case is emblematic of national patterns. An analysis of the state's institutional reforms shows how long-standing bureaucratic structures retain and reconsolidate their authority in the context of reforms that have been changing the governance of water in the postliberalization period.

India's State in the Postliberalization Era and the Case of Water

A sizable set of scholarly debates now exist on the nature of the Indian state in the postliberalization period. Scholarship on contemporary India has sought to understand and explain the nature of the changes in the political economy of the liberalizing state and to account for the weight of historical continuities that have reproduced older legacies of state power and authority despite the rhetoric of change and reform in recent decades. Such scholarship has been centered on three central theoretical frames and substantive themes that have sought to grasp the nature of state authority, institutions, and power in India. First, a large body of work has focused on understanding and assessing the shift from the developmental to the regulatory state in the postliberalization period. Such work has focused on how reforms have intersected with Indian federalism and on the potential of local state governmental policy action and innovation. A second line of inquiry has focused on the question of state capacity and effectiveness and has addressed the need for adequate institutional mechanisms to address vast socioeconomic and political challenges. The third line of inquiry has demonstrated the ways in which reforms have expanded the space for the capture of the state by private interests in ways that have intensified the inequalities and exclusions of citizenship and civil society.

Each of these frameworks captures a key dimension of the nature of state power in liberalizing India. The governance of water provides the rich terrain that allows for an understanding of each these facets of state authority and power, all of which are critical elements of the institutional framework that governs water and water-related infrastructure in India. However, while these dimensions are necessary for an understanding of water governance, they are not sufficient for a full understanding of the fault lines of the water bureaucracy. Rather, the case of water governance compels us to move a step further in order to consider how the process of reforms contains within it a set of structural contradictions that obstruct the very kinds of changes that such reforms espouse and promise. For instance, from such a perspective, the story of the remaking of the Indian state is not one of an idealized model of reforms that is shortchanged by historical legacies of state control and corruption or the capture of the state by private interests or domestic political constraints. Rather, the political economy of the liberalizing state is founded on a set of contradictions that are contained within the reforms being implemented. In the case of water governance, the effective functioning of regulatory institutions is constrained by economic pressures produced by broader sets of economic policies that are undertaken by the central and state governments and by a set of underlying centralizing imperatives that are built into processes of institutional and economic reform.

Let us consider first how the case of water governance informs an understanding of these facets of the postliberalization state. In the initial period of reforms, scholarly work focused on the positive potential of the shift from India's command-oriented developmental state to a new regulatory state (L. Rudolph and S. Rudolph 2001). This shift was in fact borne out by an increasingly active role of state governments in pursuing various policies of economic reform and drawing in private investment. In this vein, scholarship on the political economy of liberalizing India has produced in-depth comparative studies of state governments, pointing, for example, to variations between states and the increasingly visible role of some chief ministers in this capacity (Sinha 2005; Jenkins 2004).[5] In subsequent phases of the reform period, scholarly work began to call attention to the problems with the emerging regulatory state. Writing about the limits of the technocratic model of institutional transplantation that has been shaping new regulatory state structures, social scientists Navroz Dubash and Bronwen Morgan have argued that the external imposition of institutional norms and structures often operate as "shells" that conceal the domestic political and institutional practices that substantively shape policies and regulatory practices (2013, 8). Such work has pointed to the slow pace and challenges of implementing reforms and developing regulatory institutions for key sectors in the economy, such as electricity and telecom (Dubash and Morgan 2013).

The political, socioeconomic, and material nature of water is such that the governance of water illuminates such debates on the state in distinctive ways. Water involves the administration not just of a natural resource but of infrastructure. Water-related infrastructure unsettles the boundaries between the needs of traditional sectors of the economy (e.g., agriculture); sectors in the new economy, given that new industries that rely on water resources have expanded through private investment (both domestic and transnational); and the basic livelihood of citizens. The infrastructural needs of water make it a more representative case for the study of infrastructural politics than sectors that are either fully associated with the new economy (such as telecom) or those that have been more visibly associated with India's postreform model of accelerated economic growth, such as high-speed rail and highways. Furthermore, the historical weight of water-related infrastructure and institutions distinguishes the case of water from relatively new industries such as telecom, pharmaceuticals, and information technology, which are often central cases of analysis for social science research on postliberalization India. The case of water thus corrects for the methodological bias in social scientific analyses and assessments of reforms and the postliberalization state that rests on analyses of industries that are part of the new economy and that do not illuminate the historical conditions that shape the politics and political economy of state institutions. Finally, while the cross-sectoral nature of water is closer to the case of electricity, it is in many ways more enmeshed in the everyday fabric of life, as it involves a basic resource for human survival in ways that distinctively center conflicts over citizenship rights and questions of inequality within matters of governance.

In addition to its infrastructural dynamics, water cuts across territorial divisions of state administrative structures. In contrast to sectors such as electricity, telecom, or land, which have garnered much attention from political scientists writing about reforms, water unsettles the boundaries *between* states. The nature of water is such that it also unsettles the conventional methodological approach to the political economy of the liberalizing Indian state that has deployed a comparative analysis of state governments and their policies. In line with dominant methodological norms of political science, such work has adapted comparative methods to the creation of comparisons between states and state governments within India. This approach presumes discrete territorial boundaries of states and a territorialized analytical framework of the authority of state governments. The vast purview of interstate relations, negotiations, and conflicts that play out over the sharing of both water resources and water-related infrastructure exceeds the constraints of this methodological approach. In particular, the methodological terms of this comparative approach have produced significant gaps in understandings of both the nature and effects of reforms on the state and on the nature of federalism in the postreform period.

The dynamics of federalism that play out in the case of governance over water and water-related infrastructure also complicate the conventional narrative regarding the shift from the developmental to the regulatory state. The governance of water in India cuts across the various levels of federal authority in complex ways, as water has been under the purview of local, state governmental, and central governmental authority. Within India's constitutional framework, water has been under the authority of both central and state governments. While water was listed on the State List so that governance of water and water-related infrastructure was designated as a state subject, water was also simultaneously listed on the Union List of the constitution with a particular designation that both the Parliament and the Supreme Court had authority over interstate disputes.[6] The effective meaning of this constitutional designation has been that all practical matters regarding water have been under the administrative purview of state governments. The central government and Supreme Court have in practice been primarily focused on attempting to manage disputes between states, once interstate negotiations have broken down and intensified into full-scale conflicts. A decentralized framework of state authority was thus already established as a foundation for the governance of water. The effective practical authority over water supplies, distribution, and infrastructure also rested with state governments.

This did not, of course, mean that the command-oriented nature of the developmental state did not affect the control of water. The command of water resources for irrigation and the focus on agricultural productivity and the use of large-scale water infrastructure projects such as dams were central elements of the twentieth-century state developmental agendas (Frankel 2015; Prakash 1999). The technical, political, and socioeconomic approaches of state governments were shaped in significant ways by national policy frameworks and planning mechanisms of the central government. In recent years, the central government has been focused on ways of deepening national planning and central government regulatory mechanisms that can more effectively manage water resources and water-related infrastructure. For instance, in

2016, the Ministry for Water Resources, River Development and Ganga Rejuvenation put forward three major frameworks—the National Water Framework Bill, a model bill for the Conservation, Protection, Regulation and Management of Groundwater, and a report called *A 21st Century Institutional Architecture for India's Water Reforms* (Shah 2016). Meanwhile, the central government's initiation of the massive national interlinking river project embodies a reassertion of a centralized framework of state authority over rivers. Nevertheless, local state governments remain powerful actors in the governance over and control of water. The tensions inherent in this balancing and rebalancing of federal authority provide a rich terrain for an analysis of the dynamics of federalism and the tensions between centralization and decentralization in the postliberalization period.

An analysis of interstate disputes in southern India reveals the ways in which such dynamics turn on a paradox of state capacity. In the case of water, historically produced institutional weaknesses have been deepened by new pressures over water resources that stem from economic growth and the competitive framework of economic reforms that has pushed state governments to rapidly vie for private investment. Decentralization in effect marks a deep-seated form of state incapacity, as the central government has failed to provide adequate regulatory mechanisms for the management of regional and interstate water sharing. The regulatory failures of the central government are symptomatic of institutional incapacities that persist in contemporary India (Corbridge et al. 2005; Ganguly and Thompson 2017). However, such forms of state scarcity should not be conflated with processes of decentralization and privatization. This apparent weakness of the state conceals underlying centralizing processes that shape the ways in which the state commands water resources.[7] Such forms of centralization range from the assertion of local state governmental power over water and water-related infrastructure to the increasing control of metropolitan city governmental power over water to the assertion of competing centralized institutions such as the Supreme Court when the central government has failed in its regulatory role.

The mechanisms of centralization are also often underanalyzed when state incapacities are conflated with the second key feature of reforms, the project of privatization. Various forms of privatization have often taken root not as a planned policy of reforms but as a consequence of the weakness of state capacity (Kapur and Ramamurti 2002). In this context, state incapacities have meant that the state's inability to provide adequate access to necessities such as water or electricity has compelled citizens to use privatized strategies to meet their needs. Markets, in this context, are an effect of state shortcomings rather than an effective implementation of principles of economic reform. The question of state capacity is particularly significant for a careful understanding of how the principle of privatization has taken root in the context of water reforms. While the principle of privatization of water has become a key dominant global and national principle of reforms, in practice, privatization and the emergence of water markets have been symptomatic of various forms of state incapacity. In many instances, such incapacities produce illegal shadow state networks such as "water mafias."[8] The case of Tamil Nadu illustrates the ways in which private water markets emerge either through new forms of centralized city-based governance or as a substitute for state failures in providing adequate water supplies to citizens.

What is distinctive about the transformation of water governance in the postliberalization period is thus not the extent of the formal privatization of water resources. The water sector has, in fact, received relatively limited private sector investment. Indeed, processes of privatization in the water sector instead exemplify the relationships of state-based patronage, extraction, and rent-seeking that have long been a feature of the Indian state and that have been reconstituted in the postliberalization era (Bardhan 2014; Chandra 2015; Gupta 2012). Long-standing forms of state-based patronage have been reconfigured through new state–private sector relations in the postliberalization period. Such forms of patronage form a set of state practices that stem from new forms of state power that have emerged in the postliberalization period. New regulatory practices and points of state-controlled gatekeeping have replaced the old structures of the state-managed economy (Chandra 2015, 48). Such relations produce a nexus between the state on the one hand and private capital and business interests on the other (Jaffrelot, Kohli, and Murali 2019; Sinha 2005; Gupta and Sivaramakrishnan 2011). The result of this nexus between the state and private capital and the practices and patterns of rent-seeking and extraction subsequently lead to the inequities of citizenship that have been the subject of a wealth of research on the relationship between the state and civil society. Such work has provided significant insights into the ways in which citizens experience the state in a range of everyday realms (Corbridge et al. 2005; Chatterjee 2006). Citizens from less privileged groups shaped by inequities of caste, class, gender, and religion inevitably bear the brunt of a state that is captive to patronage, dominant socioeconomic interests, and graft.

While the governance of water is illustrative of the ways in which the state is embedded in relationships of patronage and private interests, a more complex set of institutional dynamics is at play. The water bureaucracy and the new regulatory institutions that have been established in the postliberalization period exemplify but are not reducible to such relationships of extraction and patronage. India's water bureaucracy consists of a complex institutional field that necessitates an examination of both variations within the bureaucracy and the complexities of the agency of state organizations and their bureaucrats. The dynamics of the captured state (whether by business or patronage politics) that are key elements of the postliberalization state coexist with state institutional structures that do attempt to engage in the kinds of regulatory practices that are invoked by idealized models of reform. An adequate understanding of how the state is compromised, captured, or incapacitated necessitates an understanding of *how* public institutions become co-opted by rent-seeking practices and organized interests.

A thick analysis of the complex and variegated institutional field of the water bureaucracy in Tamil Nadu reveals that water reforms that are designed to decentralize and improve state regulatory mechanisms in effect produce a redistribution of institutional power within the bureaucratic field of water institutions. For instance, reforms encode some city-oriented state agencies with new forms of centralized authority. Thus, some state water organizations wield considerably less power than other local state developmental institutions in ways that strain the regulatory capacity of the water bureaucracy. Meanwhile, city-based water bureaucracies gain centralized control while policies of decentralization target small towns and rural areas in ways that both reflect the political and economic weaknesses of and intensify the control of local state governmental authorities of these areas. In this process, regulatory reform is transformed into a process of differential decentralization that conceals the deeper forms of centralized state control that are embedded in frameworks of local governments. This restructuring of institutional power then enables various forms of regulatory extraction that encode relationships of power both within and between urban and rural communities.

Public Works and the Bureaucracy of Water in Tamil Nadu

Tamil Nadu (formally known as Madras State) emerged out of the British colonial Madras Presidency.[9] The Madras Presidency was the crucible of the colonial water bureaucracy, and the legacies of this colonial administration provide the context for an understanding of the historical legacies that shape water governance more broadly in contemporary India. In more recent decades, Tamil Nadu has encapsulated the complexities of India's political and economic trends. Politically, the state has been run by two regional parties: the DMK (Dravida Munnetra Kazhagam) and the AIADMK (All India Anna Dravida Munnetra Kazhagam).[10] Both parties have been part of ruling coalition governments (with the AIADMK most recently serving as part of the Bharatiya Janata Party–led coalition until the shift in power to the DMK in 2019 elections). This has meant that the state has been both substantively linked to national political trends and an illustrative case of the growing power of regional parties and coalitional politics that have been characteristic of national political trends in India. The state is also representative of major economic trends in the twenty-first century. While agriculture remains a critical part of the state's economy (with the state's Agriculture Department estimating that 70 percent of the population depends on agricultural or allied activities for their livelihood), the state is also one of the most urbanized in the country.[11] The state has actively and successfully drawn in private and global investment and has been one of the major recipients of World Bank funding in the water sector. The state has also developed a model of drawing in finance capital for infrastructure development that has been held up as a national and global model. In line with such funding and with global norms, Tamil Nadu has engaged in significant institutional restructuring of the water bureaucracy. However, the water bureaucracy has remained a powerful actor in the context of reforms, and its long-standing structures and practices are illustrative both of the older institutional frameworks that continue to shape water governance and of the new modes of centralized authority.

The pressures on water resources and the challenges of water governance in the state have also been shaped by political economic pressures stemming from geophysical attributes. Tamil Nadu is located downstream from all rivers running through it from neighboring states. This, in conjunction with repeated patterns of drought and water scarcity (that appear to be intensifying with climate change) has led to interstate conflicts and negotiations over water resources and water-related infrastructure with all three of its neighbors (Karnataka, Kerala, and Andhra Pradesh). This includes India's longest interstate dispute over the sharing of the Cauvery River between Tamil Nadu and Karnataka that has unfolded over thirty years and broken out into occasional episodes of ethnic conflict. The local dynamics of water politics in Tamil Nadu are intertwined with regional, interstate relationships and national policies and politics. The significance of the state of Tamil Nadu moves far beyond the spatial terrain of its territorial and administrative boundaries in ways that illuminate how institutional and economic reforms belie conventional narratives about federalism and decentralization.

Local water bureaucracies within states are not closed systems, and the local water bureaucracy in the state is deeply enmeshed in bureaucratic interactions with both federal and global institutions as well as within regional political and economic dynamics. Given the growing significance of local and state governments in India, an adequate understanding of the Indian state and the changes it has been undergoing in the postliberalization period necessitates a closer examination of the dynamics of public institutions. Writing about the need for such an approach, political scientists Devesh Kapur, Pratap Bhanu Mehta, and Milan Vaishnav have noted that "although much work has been done on the juridical and normative frameworks in this regard, studies on how institutions actually *work* (or not work, as is often the case) are few and far between" (2017, 2). Indeed, recent scholarship has sought to correct this significant gap through studies of key institutional arms of the state, ranging from long-standing institutions such as the Supreme Court, Parliament, and Reserve Bank to new regulatory institutions that have been set up in the context of reforms (Kapur and Mehta 2007; Kapur, Mehta, and Vaishnav 2017). The major trend in this scholarly agenda has been to focus on central institutions within India's federal structure. Consider the case of the bureaucracy. Most institutional analyses of the Indian bureaucracy have tended to focus on the centralized bureaucracy of the Indian Administrative Service (Potter 1996). However, given that economic and institutional reforms in liberalizing India have emphasized the devolution of state authority to state and local governments, there is a critical imperative to examine more closely how institutions work within local state governmental structures.

In this endeavor, I present an in-depth interpretive approach to the study of the nature and dynamics of institutional practice. The study of institutions represents a foundational line of inquiry within political science and sociology and now includes a wide range of theoretical and methodological approaches (Powell and DiMaggio 1991; Fioretos, Falleti, and Sheingate 2016). This book draws and reconfigures elements of qualitative sociological and historical approaches that have shaped the field (Schmidt 2008; Steinmo 2008). In particular, its analytical approach focuses on three key dimensions of India's water bureaucracy: (1) the historical continuities and discontinuities that shape processes of institutional formation, (2) the sociological processes of institutional restructuring and the relationship between state institutions and the sociopolitical dynamics of civil society, and (3) the agency of bureaucrats. Each of these analytical layers illustrates the reworking of state centralization in ways that constrain regulatory bureaucratic practices and intensify and produce socioeconomic inequalities.

I ground this institutionalist approach with a particular emphasis on the state's Public Works Department (PWD). The PWD represents a historic institution that played a central role in the formation of India's colonial water bureaucracy and shaped the political, economic, and infrastructural dimensions of water management in enduring ways in postcolonial India. This historical role grew out of the state administration within the Madras Presidency and became the model for the administration of water resources and public water infrastructure for the colonial state. Furthermore, the micro politics of the PWD's institutional power produced complex legacies that continue to shape postcolonial governance over water resources.

The legacy of this institutional advocacy continues on in the postcolonial context of Tamil Nadu. The colonial specificities that shape the emergence of these bureaucratic structures provide a distinctive dimension to this historical formation, as central institutional and legal structures governing water emerged through the external, extractive nature of the colonial state. Distinctive and overlapping political-economic structures place significant structural constraints on the water bureaucracy that overwhelm the regulatory capacity of bureaucratic organizations tasked with the management of increasing demands on water resources. Both the technocratic approaches of institutional reforms and the intellectual approaches of scholarly work that focuses solely on internal processes of the rule making and norms of institutions are insufficient (North 1990). The political economy of the water bureaucracy unsettles such a closed-system approach to institutions and illustrates the ways in which economic policies of reform may constrain the kinds of institutional reforms that are designed to improve the state's capacity to govern effectively.

A case study of the PWD thus provides a distinctive lens on the historical legacies that have shaped India's water bureaucracy. Such a historical institutionalist perspective allows for a delineation of the continuities and discontinuities between the colonial and developmentalist state that are reconfigured again in distinctive ways in the postliberalization period. The choice of the PWD as a core case of analysis for this study also rests on the continued significance of this bureaucratic organization in the postindependence period in Tamil Nadu. In contemporary Tamil Nadu, the Public Works Department has retained control over irrigation as well as over the regulation and storage of water and maintenance of water bodies. In keeping with the historical weight of its institutional authority, it is the only such department in the country with control over irrigation. This preservation of authority has meant that the PWD has remained a leading institutional actor within the water bureaucracy in Tamil Nadu.

The significance of the PWD also rests on the way in which it is emblematic of long-standing state institutions that have shaped developmental policy in contemporary India. While Tamil Nadu's PWD is unique in retaining water within its purview, it is emblematic of the kind of bureaucratic organizations that oversee both water resources and infrastructure. It is representative of organizations that have governed rural drinking water supplies in India, such as the Public Health and Engineering Departments (S. Singh 2016). It is also representative of the kinds organizations that have governed infrastructure more broadly and that have often been viewed as the face of what is now commonly known as the "engineer-contractor-politician nexus" in India. Consider former Indian Administrative Service officer Gajendra Haldea's critical discussion of major infrastructure projects such as the redevelopment of the Delhi airport and the Games Village for the Commonwealth Games in India. Extravagant airport complexes and megaevents such as the Commonwealth Games have become some of the iconic spatial-symbolic stages that nation-states in comparative contexts have invested in so as to project a vision of their standing within the era of reform and globalization. Writing about the scams and mismanagement of such infrastructural projects in Delhi, Haldea describes at length an "engineer-contractor-politician nexus," where, he notes, in a conventional PWD-style contract, bids are invited for the unit rates payable in respect of each item of work. The government engineer measures each unit of work and makes a running payment for the work done. In this process, the costs of additional quantities and new items are also paid by the government. Moreover, delays on this account are borne by the government and the contractor is compensated for inflation during the construction period. In effect, this is like an open contract which offers enormous opportunities for time and cost overruns as well as for corruption. The engineer and the contractor have little incentive to complete the work in time and within the estimated costs. (2011, 97)

Such dynamics represent a familiar story of infrastructural politics in India and form a well-known component of contemporary media and public discourses on state corruption (Heller, Mukhopadhyay, and Walton 2019). What is of significance is the centrality of this typology of state institutions in various localities (and at both the local and the central levels) that continue to play a critical role in the postliberalization period but remain an understudied arena of the institutional infrastructure of the state.

If the PWD has continued to play a central role in water governance in Tamil Nadu, its institutional monopoly has also been weakened by various phases of institutional reform. While the PWD's role in managing irrigation and water infrastructure has continued, institutional reform and new forms of urban governance have also increased the power of urban water supply and urban development bureaucracies of the major metropolitan city of Chennai (such as the major water utility, Metrowater). Tamil Nadu's water bureaucracy exemplifies how global and national principles of institutional reform are implemented at the local level, their effectiveness, and their implications. New processes of centralization are taking shape in the context of liberalizing India through shifts of centralized power within institutional fields rather than through substantive forms of decentralization. The broad dynamics of federalism, decentralization, and state capacity can be adequately addressed only by delving into the very institutions that are meant to be the arenas of the state governments that are now designated as the new crucibles for economic reform, growth, and governance.

Finally, this study of institutional restructuring seeks to deepen an understanding of the structural and institutional constraints on regulatory reforms by incorporating a fuller account of the agency of bureaucrats within the

Public Works Department. The Indian bureaucrat is perhaps one of the most heavily typecast figures in contemporary India. Some of the early romanticized characterizations of the Indian bureaucrat as the capable, professional embodiment of the Indian Administrative Service have long since given way to two typologies. The weightiest discourse on the Indian bureaucrat has been centered on a character prone to corruption and abuse. A wide range of discourses across the ideological spectrum within public and media narratives in India, academic work concerned with equality and social justice, and pro-reforms rhetoric on the inefficiencies of the state converge on a narrative of the bureaucrat as the central obstacle to economic development, progress, and equity. A second typology has sought to recover the agency of the bureaucrat as a rational actor that has navigated the structural and cultural constraints of India's centralized state (Sinha 2011). Both typologies present key components of the nature of bureaucratic complicity and agency that are an important dimension of the agency of water bureaucrats. However, these typologies also present the bureaucrat as a static, homogenized individual driven by self-interest—whether of monetary gain or vested institutional interests in particular outcomes. What is missing in this context is an understanding of the bureaucrat as a complex subject whose motives and agency may move beyond these familiar typologies.

This book develops an approach to institutions that incorporates this sense of complexity by treating the bureaucrat as a subject of history. There is of course a now extensive scholarship that has sought to give subaltern subjects a place in history with the kind of complex subjectivity and agency that was once the preserve of elite-centered histories. Yet, fraught subjects such as bureaucrats, who often occupy contradictory socioeconomic and constrained political positions, do not conform either to ideological visions of the bureaucrat responsible for the implementation of models and practices of development that may further marginalize the poor, on the one hand, or to stereotypes of the bureaucrat as the emblem of the bloated state that needs to wither away in service of economic reforms and growth, on the other. To that end, this study weaves in a more complex sense of the agency and subjectivity of bureaucrats. This includes both the spaces in which bureaucrats attempt to perform their regulatory functions within the constraints of political pressure and structural forces and the moments in which individual bureaucrats develop and put into practice their own sense of ethical agency.

Methodology and Interdisciplinary Framework

This study is based on an interdisciplinary social scientific approach to the study of the state. While framed by an engagement with political scientific and sociological debates on institutional reforms, the methodology draws on interdisciplinary interpretative methods. In particular, it draws on an adaptation of Geertzian methods to provide a "thick description" of the historical, institutional, and political-economic underpinnings of the water bureaucracy. This mode of interpretative methods is in line with political scientific approaches that have reconfigured such methods in order to address structural and systemic explanations (Wedeen 2002). Such social scientific approaches are different from anthropological ethnographies that delve into a detailed description of a single microinstitution.

My focus on the systemic continuities and changes in the institutional dynamics of the water bureaucracy engages with an analysis of changes at the national, state, and local levels of the water bureaucracy. Such an approach is concerned with systemic changes that are shaped by economic and institutional reforms. The empirical research that I draw on in this analysis consists of a wide range of archival research, including historical, local, national, and global policy materials, as well as qualitative interviews and field site visits that I conducted in Chennai between 2016 and 2018. The city of Chennai is a central site of analysis both because it is the site of power for the central water institutions and because it is the locus for understanding the ways in which city-centered urban governance has become the key node for the processes of state centralization that this book is analyzing. As one of India's major metropolitan cities, it provides a critical and representative site of broader national patterns regarding water governance.

The empirical research on the PWD, the central case of the book, is based on exhaustive archival work that I conducted at the organizational archives of the PWD. These materials included colonial documents, policy studies, and internal PWD documents and reports that ranged from the 1950s to the present. These materials also included the autobiographical writings and professional records of the late A. Mohanakrishnan, one of the central figures in the PWD, whose career spanned the period from 1947 to 2012. This service, in addition to his role as the chairman of Tamil Nadu's Cauvery Technical Cell (responsible for negotiations with Karnataka), provides a unique window into the complex nature of bureaucratic agency. While the heart of the research draws on these archival materials, the study also utilizes qualitative interviews with PWD employees. Finally, the study also draws on field site visits and interviews at related water bureaucracies and independent agencies, including Metrowater, Chennai's major water utility (formally known as the Chennai Metropolitan Water Supply and Sewerage Board), the Tamil Nadu Urban Infrastructure Financial Services Limited, and the Smart Cities Mission agency associated with the Chennai Municipal Corporation (Greater Chennai Corporation).

While this book addresses institutional reforms within India, it also informs broader comparative and global debates on the governance of water. In the water sector, global norms have either directly or indirectly shaped national and local policies, discourses, and bureaucratic practices (Asthana 2009). While a range of transnational actors shapes global norms and practices, comparative scholarship has well documented the central role of the World Bank in shaping ideational frameworks and policies in comparative contexts (Bakker 2010; Morgan 2011). The role of the World Bank in India has been more complex than a presumed imposition of global norms of privatization might suggest. Such global processes impact state authority through Bank policies and projects in India. The Indian state has always been a strong actor, including in its historical engagement with the World Bank. The centrality of the state is evident also in World Bank policy and project frameworks—the role of centralizing state authority is present in subtle ways even in global frameworks that are rhetorically identified with the principles of privatization and decentralization. Given the expansive scholarship on the privatization of water in comparative contexts, this often subtle but sustained significance of the state may appear to make India an exceptional case. However, a closer analysis of global trends indicates that the Indian case in fact points to a deeper and often understudied role of the state in the water sector in comparative contexts.

Inequality, Urban Governance, and the Politics of Water in Comparative Perspective

The case of water reforms in India provides important insights for a comparative and transnational understanding of the governance of water. Given the significance of the global push toward privatization that emerged in the 1990s, comparative and transnational scholarship has often understandably focused on the impact of programs of privatization implemented through various sets of policy reforms (Chng 2008; Hall, Lobina, and de la Motte 2005; Harris and Roa-García 2013; Ioris 2012; Schnitzler 2008). Such scholarship has examined both the effects of privatization and the protest movements that have focused on privatization and, more generally, on "neoliberalism" in a range of cases in Latin America, Southeast Asia, and Africa (Bakker 2010; Morgan 2011; McDonald and Ruiters 2004). Indeed, within India, critics of economic reforms have also focused on the threat of privatization.[12] However, in the post–Washington Consensus era, state accountability has emerged as a vital foundation for global policy frameworks. Furthermore, both intense domestic political opposition to privatization and the complexities and in some cases uncertainty of investment returns have dampened private investment in the water sector. By the late 1990s, this led to a partial retreat of the private sector (Bakker 2010). While this book takes processes of privatization seriously, the Indian case points to a deeper set of implications that are not adequately captured by debates on the advance or retreat of privatization. The Indian case points to the necessity for understanding the role of the state in controlling and distributing water resources, even in a context where privatization is taking place. Tamil Nadu is a significant case in the Indian context precisely because state centralization is taking place in a context where the local state government has embraced global norms of decentralization and privatization and has been a leading actor that has specifically worked with the World Bank.

The Indian case provides a set of policy and political dynamics that at first glance appear to depart from the comparative cases in other late-industrializing countries that have implemented global norms. In contrast to other cases, India's water sector has been characterized by the relatively late implementation of principles of privatization and state restructuring and the limited extent of private investment. However, these specificities are now representative of comparative trends that have seen a retreat in privatization and the post–Washington Consensus focus on the state. At one level, the centrality of the state provides a counterpoint to the past focus on the centrality of privatization in shaping both changes in the water sector and the creation and intensification of inequality. Despite the extensive rhetoric on privatization (by both proponents and critics), by the early twenty-first century, only about 5 percent of the world's population was being served by the formal private sector (Budds and McGranahan 2003, 88). Indeed, recent scholarship has begun to point to the need for a more intensive analysis of water governance (Akhmouch 2012) and has called attention to the ways in which comparative cases illustrate that the key principles of privatization and decentralization have not unfolded according to the idealized norms of the global model.

Consider, for instance, the case of Chile, one of the strongest and earliest models of the privatization of water, in which water rights were treated "not merely as private property, but also as a fully marketable commodity" (Bauer 2010, 46). Nevertheless, even in this strong model, the state's regulatory framework has provided a critical, if contradictory, foundation for privatization. Madeline Baer, for instance, has argued that "Chile's successful water sector is the result of the creation of an efficient public water sector prior to privatization and of the state's capacity to govern the sector to mitigate the negative effects of privatization" (2014, 142). This success has concealed underlying problems of governance. The regulatory framework has provided economic benefits by encouraging private investment for agricultural, urban, and industrial uses but has failed to handle complex questions of governance involving competing demands on water resources, questions of equity of access to water, and environmental issues (Bauer 2010, 46). These differing assessments point to the role of the state's capacity and regulatory framework in understanding the relative successes and failures of the Chilean model.

The Indian case allows us to understand the implications of this trend in which principles of privatization and reform are embedded within state-led endeavors, policies, and processes. Such an approach enables a shift from a static understanding of ideologically driven conceptions of privatization that both critics and proponents of liberalization often deploy. As social scientist Karen Bakker has aptly noted, such ideologically driven policy debates "rely on the assumptions of utilitarian liberalism, in which the distinction between public and private equates with that between governmental and nongovernmental" (2010, 29). Instead, processes of institutional reform transform the boundaries and meanings of the categories of "public" and "private" that reinforce state power and rework societal inequalities and exclusions in more complex ways that exceed a more static conception of privatization (Bear and Mathur 2015). Such changing conceptions undergird the kinds of inequalities and exclusions that both constitute and are reconstituted by the remaking of state power in liberalizing India. However, this book is not simply concerned with illustrating that the state still matters. Rather, its focus is on the ways in which water reforms in India provide a case for understanding the distinctive form of centralizing state power that is remade in the postliberalization period.

Consider, for instance, the ways in which policies of decentralization have been implemented in the water sector in comparative contexts. In Cape Town, South Africa, a decentralized policy for water provision was shaped by the power of city-level policy makers and bureaucrats that intensified the top-down approach of the city (Yates and Harris 2018, 78). These subtle centralizing trends are evident in a range of cases where policies of decentralization have produced new forms of concentrated authority over water resources by local elites, bureaucrats, and government officials. In the São Francisco River Basin in Brazil, water reforms reinforced elite control over water resources through an underlying pattern of the elite capture of regional bureaucratic power (de Freitas 2015, 298). Similar patterns of continued state and elite control have been documented in a wide array of national contexts, including Colombia, Peru, Kenya, and Turkey (Guerrero, Furlong, and Arias 2015; Ioris 2012; Islar and Boda 2014; Kemerink et al. 2016; Swyngedouw 2004). In the case of Colombia, "administrative decentralization took place through the devolution of authority to municipalities. This new authority, however, was rapidly withdrawn from smaller localities, semi-'recentralizing' it to departments and regional bodies" (Guerrero, Furlong, and Arias 2015, 173). Such fine-grained comparative work underlines the significance of adequately understanding how such processes of centralization play out in the context of water reforms. An argument of Indian exceptionalism that suggests reforms are simply blocked by the particular historical conditions of Indian political and institutional life does not adequately account for the ways in which the dynamics of state centralization are in fact built into the very policy, institutional, and political-economic frameworks of reform that claim to rest on idealized norms of decentralization and privatization.

Such processes of state centralization shape the extraction and distribution of water resources in the context of increasing pressures produced by economic growth, urbanization, and the intensifying effects of climate change. This book analyzes the historically produced institutional practices that create inequalities in the allocation of water resources. What is at stake in this endeavor is an understanding of *how* effective institutional responses to the challenges of inclusive and accountable governance are constrained and foreclosed. Such an understanding foregrounds the ways in which local governmental agencies and their bureaucratic actors play crucial—and understudied—roles in responding to the macro national and global economic and environmental crises of the contemporary world.

CHAPTER 1

The Historical Formation of India's Water Bureaucracy

WATER INSTITUTIONS HAVE BEEN FORGED THROUGH A COMPLEX set of historical processes in contemporary India. Contemporary water institutions have been shaped and constrained by patterns of institutional practice that were emblematic of the politics and political economy of both the colonial state and the early decades of the developmental state in independent India. In particular, the formation of India's water bureaucracy was shaped in significant ways by the institutional development of the Public Works Department that was established in colonial India. Like many of India's bureaucratic structures, these institutional legacies were later recast within India's newly independent state.[1] The institutional structures of India's water bureaucracy were forged through colonial state developmental policies that were geared toward large-scale irrigation projects designed to maximize state revenue (Goswami 2004; Ludden 1979; Mollinga 2003; Mosse 2003). However, the colonial water bureaucracy was also marked by distinctive structures, interests, and practices that were more varied than the centralized colonial state's singular interest in revenue extraction. While land remained the foundation for the exercise of state power (and the collection of revenue), water became a fraught terrain in which colonial state power was exercised, institutional battles within the colonial state apparatus were fought, and civil (and later nationalist) resistance arose. This terrain was framed by the colonial state's debates, policies, and conceptions of "public works." Public infrastructure, in particular, became the site for the exercise of state power through a central arm of the colonial state, the Public Works Department. The Public Works Department's advocacy of water infrastructure and model of irrigation development in turn shaped colonial state power in fundamental ways.

The complex political and economic configurations of the colonial state are vividly encapsulated in Lieutenant Arthur Cotton's depiction of "public works" as the foundational infrastructure of the colonial state. Thus, he argued,

> It may be said, that Public Works are very secondary matters, and that after providing for the protection of the country, Civil and Judicial affairs are after all the great things to be attended to. The question of Public Works is however in reality a fundamental point; for upon it and upon it mainly depends the capability of the country to supply funds for every purpose both Military and Civil. Without Public Works, the country must remain sunk in poverty and ignorance, for funds cannot possibly be produced from the country itself to provide for what is necessary to elevate and improve the state of the people; whereas, with Public Works, the most abundant funds can be obtained for any purpose. (1854, 61).

Cotton, one of the earliest and most vocal advocates of public works, was particularly invested in state investment in large-scale irrigation works. His conception of such works as integral to every military and civil purpose reflected a complex understanding of the political and economic power of infrastructural investment by the colonial state. Public works of irrigation, in this vision, were at the heart of all aspects of state power. As a source of revenue, they provided the foundational financial underpinnings of colonial state power. As a means for the presumed social improvement of "the people," they provided a critical basis for state ideological justification for colonial rule; for instance, such works were a symbol of technological prowess and improvement that embodied the ideological rationale for the state's civilizing project that would rescue a country otherwise "sunk in poverty and ignorance."

While these facets are familiar dimensions of colonial state power, Cotton's vigorous advancement of public works as the foundation of the judicial, civil, and military dimensions of state power also provides a window to a deeper understanding of an emerging nexus between bureaucratic power and infrastructural politics in the colonial period. In this story, Cotton's characterization of the power of public works is not simply an empirical description of the workings of colonial state power but a project of institutional advocacy in a stratified and contested bureaucratic apparatus within the colonial state. Cotton's forceful argument about the power of infrastructural projects of public works was in fact designed to consolidate and expand the institutional power of the newly formed Department of Public Works within the broader bureaucratic field of the colonial state apparatus.

The Public Works Department (PWD) grew into one of the central colonial state institutions that oversaw state investment and management of all infrastructural investments, including the key sectors of railways, roads, ports, and irrigation. However, Cotton's institutional advocacy was more targeted than this general portfolio of colonial state infrastructural power. As an irrigation engineer, his career was marked by an ardent and persistent campaign to promote water-based infrastructure designed both to increase agricultural productivity and to protect the interests of the irrigation bureaucracy within the Public Works Department. At one level, the colonial state's expanding investment in and imposition of new models of irrigation significantly shaped methods of agricultural water usage. However, the process of reshaping water usage was not merely a reflection of the dominant discursive norms emerging in fields such as engineering or of the centralized interests of the colonial state in expanding its sources of revenue through increased agricultural productivity. Rather, the increasingly powerful and independent institutional interests of the water bureaucracy within the Public Works Department became a critical factor in shaping the control and distribution of water resources.

The case of the Madras Presidency is particularly salient because it was the crucible for Cotton's advocacy for colonial models of irrigation that drew on his engineering training. Cotton's work in the Madras Presidency provided the basis for his broader institutional advocacy for the Public Works Department's irrigation bureaucracy and became a template for other parts of the Indian subcontinent. Such micro politics of the PWD's colonial institutional power produced complex legacies that continue to shape postcolonial governance over water resources. The historical institutional legacies of the colonial state encompass both the broad patterns that shaped the political economy of the colonial institutional infrastructure of the PWD and the less visible micro practices of local bureaucratic organizational approaches, cultures, and practices. For instance, the PWD's role within water-sharing disputes and negotiations between the Madras Presidency and neighboring princely states has shaped postcolonial water-sharing arrangements in southern India. Meanwhile, the legacy of this institutional advocacy continues on in the postcolonial context of Tamil Nadu. As the only state where the PWD has retained control over the governance of irrigation water, this focus on the Madras Presidency and on the state of Tamil Nadu that emerged from this presidency allows for a sustained qualitative analysis that can trace the historical continuities and changes in this institutional framework from the colonial era to the current postliberalization period.

The historical weight of the PWD's colonial-era institutional practices was reinscribed in complex ways within the bureaucratic structures in postcolonial India. However, while India's postcolonial state was shaped in significant ways by these historical patterns, this newly independent state was also tasked with breaking from many of the legacies of the colonial state. The politics of water was at the center of the developmental state that shaped the trajectory of India's political economy path. Whether through Nehru's oft-cited metaphorical characterization of dams as the "temples of modern India" or through the agricultural policies of the Green Revolution, the harnessing of water resources rested at the heart of India's modernist vision of the developmental state. This vision rested on the state's drive to harness water for large-scale developmental projects, such as large dams, and produced a sustained and intensified use of water resources through modernized and expanded irrigation infrastructure and the exploitation of groundwater resources. However, beneath the image of this singular, overarching "modernist state" lay the complex and powerful water bureaucracy that directed, implemented, and reinforced this model of water governance.

The interventionist and extractive dimensions of the water bureaucracy were reshaped and intensified by the developmental goals of the postindependence state. While the centralizing impetus of the colonial state was reoriented toward new developmental goals, this was accompanied by weak and delayed regulatory mechanisms. India's constitutional framework placed the governance of water within a decentralized framework that gave local state governments primary authority over water resources and infrastructure. The central government's emphasis on the instrumental, extractive use of water without establishing national regulatory institutional frameworks expanded the space for local state governmental authority. National regulatory institutional weaknesses and gaps were filled by local state governments. This in turn intensified bureaucratic centralization at the local level. While the goals of the state were significantly different from colonial governmental interests, the centralizing, extractive nature of local state water institutions was both consolidated and intensified by the interventionist developmental state in independent India. This consolidation of state intervention was shaped as much by the centralization of state control at the local level as it was by familiar images of India's planned developmental state.

Public Works and the Emerging Water Bureaucracy in the Madras Presidency

Large infrastructure projects represented central and visible elements of state power in colonial India. Such works, ranging from building construction to communications to irrigation projects, were both the means that the colonial state used to consolidate and exercise power and the symbolic embodiment of colonial rule (Goswami 2004, 47). However, a closer examination of irrigation works illustrates that this cohesive framing of the material-symbolic power of the colonial state, in practice, concealed a messier, contested institutional history. Public works of irrigation did not emerge as an inevitable technology of state power. Rather, the emergence of such technologies of colonial state power was the product of institutional interests, bureaucratic lobbying, and intrainstitutional bureaucratic competition. While public works of irrigation eventually became a key component of the political economy of the colonial state, it was the product of a set of microinstitutional dynamics that emerged within the Madras Presidency. The construction of major public works of irrigation in the Madras Presidency in turn shaped broader colonial discourses and policies in the management of water resources in the Indian subcontinent.

Early advocates of public works of irrigation based their arguments on the central ground of state interests—the ability to generate expanding revenue from land. As early as 1856, proponents of colonial state investment in irrigation pointed to the early economic profitability and productivity of such works initiated by the Madras government. As one such advocate argued, "The irrigation works constructed by the Government in the Madras presidency—which may be taken as a type of the general character of

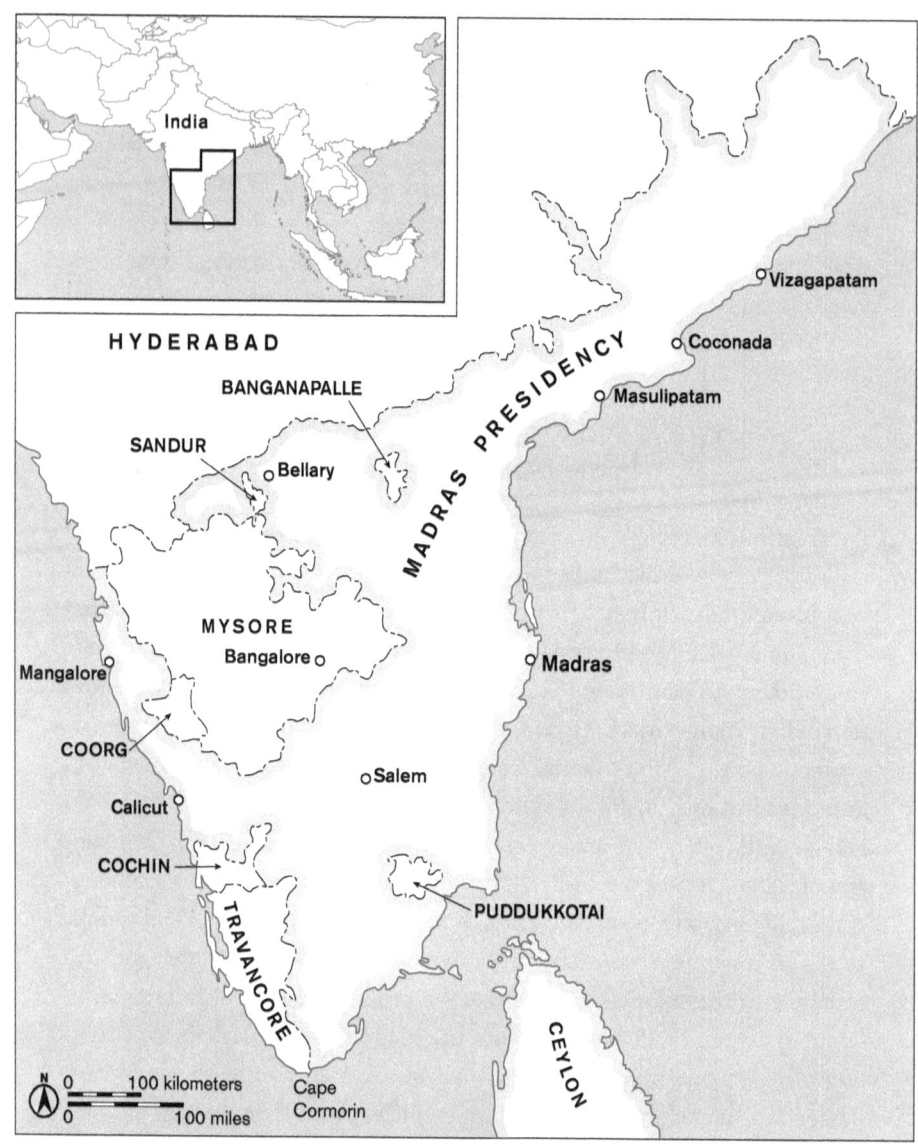

MAP 1.1. The Madras Presidency in Colonial India, showing the administrative territory under British colonial rule in southern India. Adapted from public source material: Edgar Thurston, *The Madras Presidency with Mysore, Coorg and Associated States* (Cambridge: Cambridge University Press, 1913).

irrigation works suitable for India—have so much added to the revenue of the locality in which they are situated, as to have brought back the Government a return equal to a dividend of 70 per cent per annum upon the capital invested in them" (Bourne 1856, 9). The relationship between irrigation works and revenue collection became a naturalized tenet of colonial state policy. However, this belief that later undergirded the state's primary focus on irrigation as a source of revenue collection (Ludden 1979; Mosse 2003) was not a natural or self-evident truism but a product of the persistent and arduous institutional advocacy from state bureaucrats within the Public Works Department (PWD).

Governor General Dalhousie authorized the formation of the Public Works Department, the major bureaucratic institution that was to oversee such works in 1854. Works that were previously under the purview of the British Military Board were now transferred to this new institution. In the context of these institutional changes, proponents of irrigation works represented an emerging and increasingly powerful bureaucratic lobby within the Public Works Department. Proponents of water-based infrastructure and investment actively competed with bureaucrats in favor of railways. In this endeavor, the newly emerging colonial water bureaucracy specifically made claims tied to the public good. Arguing that "the railway companies are powerful and rich," John Bourne insisted that it was investment in water for both irrigation and navigation that was in the best interest of the colonial government (1856, 28). The competition over the choice between railways and water was an intrainstitutional debate within the PWD on which forms of infrastructural investment would be most profitable and in the interest of the state. The emerging water bureaucrats insisted, for instance, that "a bad water-communication will carry more cheaply than a good railway" (Bourne 1856, 28).

While the water bureaucrats would, of course, prove unsuccessful in displacing the power of the railways in India, they were able to construct a powerful foundation for their institutional authority when it came to the promotion of large-scale irrigation works. This case rested on the creation of a conception of public interest that wove together the economic and political interests of the state, the welfare of "the people" of India, and the technical expertise and authority of the emerging water bureaucracy. Consider, for instance, how such connections are created through the elaborate arguments of Arthur Cotton, the leading proponent of this vision. Cotton's case for public works rested on a complicated approach to the question of revenue. At one level, his writings in the mid-nineteenth century represented an acidic attack on the British government's narrow focus on revenue collection. As he put it, "They still cling to the old idea, that if only the subject of *collecting the Revenue*, be attended to, everything else will follow as a matter of course" (1854, 10). Cotton's reframing of the question of revenue was not aimed at reducing or refraining from the collection of revenue from the people but was intended to ask instead "how to enable them to pay it?" (12). This reframing of the question formed the basis of the new model of public works of irrigation that shaped colonial polices toward the development of "the people" of India. The "welfare" of the people in effect became bound up with the state's interests in revenue collection. The foundation for this new relationship between welfare and revenue was state investment in public works of irrigation. To this end, Cotton marshaled evidence designed to prove this relationship. Drawing on the model of the district of Tanjore in the Madras Presidency, he argued that in this case steady investment had produced "unvarying success, raising Revenue from 30 to 50 lacks (£500,000) a year" (19). This example, for Cotton, was representative of "so many proofs of the immense losses resulting from the neglect of Public Works and the enormous profit derived from the execution of them" (19).

If Cotton's arguments were centered on revenue and welfare, his impassioned case for public works was also inextricably linked with a set of underlying institutional interests. Most of his arguments were focused on trenchant criticisms of the British civil service, which he argued was focused solely on "the Revenue Settlement" and the "mode of collecting" (47). Cotton's arguments soon become entrenched knowledge. By 1860, the Madras Irrigation and Canal Company pointed to evidence of "large returns realized by the Government" and noted that the "immense political and social importance, and the highly remunerative character of Works of Irrigation in India, are now so well understood, that it is unnecessary in a prospectus to advance proofs" (EIICC 1860). This prospectus for private investment, prepared under the orders of the colonial government with Cotton as "the highest authority in such matters" (7), expressed confidence that such private investment would "of necessity, yield considerable returns" (10).

This new model of financially productive public works that claimed to serve both the interests of the state and the welfare of the people was the foundation for a new state bureaucratic apparatus. Cotton was focused not simply on increasing state investment but on the creation of a new institutional structure for this avenue of capital investment. In this institutional model, the Public Works Department would "keep a complete and distinct set of accounts of its own, and everything spent by it should be divided under two heads, 1st current expenditure on superintendence and necessary repairs … [and 2nd,] new works including improvements … to obtain a tolerable estimate of the actual total returns of all new works" (67). The new department was aimed at correcting the existing situation, where "Public Works have been almost entirely neglected throughout India" (272) in large part, according to Cotton, because of a lethargic process where requests for expenditure were channeled through a long institutional channel of authorization that ran through the collector, the Board of Revenue, the Local Council, the Supreme Council, and finally the approval of two more boards within England (274). The new Public Works Department would, in effect, regulate the relationship between the colonial state, revenue, and the needs of the populace and provide a new irrigation bureaucracy in the Madras Presidency by the 1870s (Ludden 1979, 358). This process of institutionalization produced a model of the "public" interest that rested on the connection between capital investment, state interests, and the socioeconomic well-being of the people.

The power of the emerging water bureaucracy's institutional lobbying can be discerned by moving away from the presumption that the expansion of irrigation works was a natural outcome of the state's modernist conceptions of the economy. Consider, for instance, the salience of a model of public works emerging in the Madras Presidency becoming a generalized technology of state power. Within the framework of the colonial state, the Madras Presidency itself was a tertiary rung below the colonial governance structure in London and Calcutta. To the upper tiers of the colonial state, the Madras Presidency was merely a source of income and "appeared a bottomless purse which could be looted whenever need arose" (Washbrook 1976, 24). In his rich analysis of the institutional complexities of colonial rule, historian David Washbrook further argues that given that the upper tiers of the colonial state could both veto legislation and demand revenue, the power of the Madras Presidency rested with local bureaucratic authority exercised in realms such as "the assessment of land revenue, the distribution of government jobs and contracts and the construction of irrigation works" (25). Within this context, the institutional lobbying of water bureaucrats within the

Madras Presidency represented an attempt to push against the political constraints of the hierarchical and stratified structure of the colonial state. As evidence of the success of this strategy, we see that this model of public water infrastructure soon extended beyond the Madras Presidency. The colonial state's model of flood control in Orissa in eastern India was developed based on Cotton's advocacy of an elaborate canal system for mass river transport that would compete with the railways (D'Souza 2006, 133). The dominant institutional models of water management that emerged from the Madras Presidency overrode independent investigation into the causes of flooding (D'Souza 2006). This dominant colonial PWD model of water management thus had significant effects on the governance and management of water resources well beyond southern India.

In the case of the Madras Presidency, the expanding power of the water bureaucracy created significant shifts in the political and economic life of agrarian communities. For instance, the emergence of this regional bureaucracy reconstituted community-state relationships in complex ways (Mosse 2003). This reconstitution was not simply a product of a simple form of state centralization and the fact that water resources were linked to the overarching economic and political power of the state. Rather, shifts under colonial rule produced deeper changes in the nature of political accountability and authority of the state.[2] Consider, for instance, the changes in management of the vast network of tanks in colonial Tamil areas of the Madras Presidency that ensued from the steady bureaucratization of water resources. Irrigation (in what would later become the state of Tamil Nadu) had been based on an engineering model of water tanks. River water was diverted into tanks through channels that had been dug, and in cases where water needed to be diverted between villages, a system of cascading tanks would divert excess water. The gradual erosion of proprietary control of zamindars over tanks under colonialism produced the systemic deterioration of the tank system. The colonial model of revenue collection transformed preexisting logics of royal patronage so that, as anthropologist David Mosse has argued, "it was not (as it had been) a means to create autonomous nodes of investment in irrigation, but rather a device to generate cash flows to the zamindars' offices in the form of banker's credit or commandeered temple funds without the obligation to invest in public goods" (2003, 79). Such processes were compounded by the increased insecurity over land tenure in the context of changes in the colonial economy that further reduced incentives for local investment in tank maintenance (Mosse 2003). Public works of irrigation thus became intricate networks for the exercise of bureaucratic power, politics, and patronage (Vasavi 1999). Bureaucratic corruption became widespread as the power of irrigation bureaucrats expanded, and it later produced populist resistance to the PWD (Washbrook 1976, 180).

These historical legacies of colonial institutional practices have had enduring effects in the postcolonial period. In the case of the emerging water bureaucracy in the Madras Presidency, the new colonial Public Works Department was being formed by a set of dispositions that shaped institutional practices in durable ways. The result was the creation of a distinct set of institutional customs that influenced the agency and outlook of this new bureaucratic authority. The early institutional lobbying for the PWD, the political claims for financial autonomy, the distinctive emerging institutional discourses on the meaning of the "publicness" of public works of irrigation, the internal competition with the railways and revenue wings of the state, and the expansion of local bureaucratic power all created the components of a distinct form of the bureaucratic agency and culture that later continued to shape the management of water resources in postcolonial India. The institutional field of the water bureaucracy in colonial India had emerged as a discrete entity that was focused on its own self-reproduction and political and economic power through the pursuit of public works of irrigation.

Bureaucratic Agency, Culture, and the Interests of the Colonial Water Bureaucracy

The emerging water bureaucracy in the Madras Presidency was defined by a distinctive bureaucratic culture that became deeply embedded in the state's approach to governance over water resources. A defining element of this bureaucratic culture was the emergence of a fractured institutional structure that produced strong interinstitutional rivalries between competing bureaucracies. We have already seen that the irrigation wing of the PWD jostled in competition with the powerful railway wing from the inception of the PWD. While the proponents of works of irrigation successfully carved out their institutional territory and shaped colonial policies in significant ways, the mode of bureaucratic competition inherent in this early contestation remained a significant feature of the water bureaucracy, particularly as the irrigation wing of the PWD gained more authority. A key element of these interinstitutional struggles stemmed from the relationship between the PWD and the Revenue Department—one of the major arms of colonial state power, given the centrality of revenue collection for colonial rule. Prior to the PWD's autonomy, the administration of public works in the Madras Presidency was fragmented under the Revenue Department, the superintendent of roads, and the Military Board (MP 1852). In the case of irrigation, this meant that English collectors "assumed the charge of works of irrigation with that of collecting revenue" (MP 1852). The reorganization of this administration, in response to the institutional lobbying put forth by Cotton and other advocates of the water bureaucracy, placed the PWD in charge of works of irrigation. The result was the beginning of a long history of interinstitutional competition over the control and governance of water resources and water-related infrastructure.

One of the central bureaucratic cleavages was between the technically oriented bureaucrats in the PWD, who were tasked with the design of new irrigation works, and the administratively oriented collectors within the Revenue Department. As early as 1856, internal correspondence reveals problems with the communication between civil engineers of the PWD and the Revenue Department (Grant 1857). In such cases, engineers at the district level would develop proposals for irrigation works without including the perspective of revenue collectors. The chief engineer and head of the PWD reproached such officers for the "neglect" of "the not unfrequent omission of all reference to the Revenue authorities in the several projects of new works, recently submitted for entry in the Budget of 1856–57" (Maskell 1856, 28). The underlying fissures reflected new divisions between the respective executive and administrative functions of the two departments. In the new institutional division of labor within the colonial apparatus, the engineering cadres of the PWD were marked by their ability to bring "science and professional experience to bear upon the performance of works requiring skill and practice." Meanwhile, the collectors of the Revenue Department were to "henceforth assume the more befitting and appropriate position of general administrators of the province, and . . . become the immediate referees of Government respecting the success and effects of the public works carried on by the Professional Department" (248). The complicated relationship between the two departments was underlined by the fact that while the aim of the establishment of the PWD was to create an autonomous institution, estimates for new works and improvements of existing works were to be countersigned by the Revenue Department collector (147). The entangled nature of this emerging relationship between the two organizations reflected the teething pains of working out the increasing political power of claims to professional expertise and the primacy of the aims of revenue collection of the colonial state. Early documents of the newly formed Public Works Department reveal the initial attempts at smoothing out the working relationship with the Revenue Department. However, the very call for "unity" between the engineers and the collectors made by the head of the PWD underscores the cleavages between the two organizations (29).

The colonial state attempted to manage this relationship between the two institutions through the creation of rules of engagement that set in place the lengthy tracks of paper-based practices of communication and reporting and that are now a notorious sign of India's bureaucracy. The Board of Revenue was, for instance, to "direct Heads of districts to submit yearly with the settlement report, a report upon the progress of improvement in their districts, embodying in it the purport of the monthly lists to be furnished by the Civil Engineers" (45). Collectors were responsible for tracking spending on public works "and its effect on the revenue of Government and the welfare of the people" (46). By the turn of the century, the state was still wrestling with the division of labor between the PWD and Revenue Department. A report on the colonial government's Tank Restoration Scheme in the Madras Presidency argued that the restoration of small tanks that were irrigating under two hundred acres "be placed under the control of the Revenue and not the Public Works department" (MP 1902, 131). Such tensions were not a mere form of territorial jockeying between bureaucrats. They were an integral element of problems with the colonial model of irrigation development that had profound effects on rural society in India.

The *Report of the Indian Irrigation Commission* highlighted the problems of such fractured administration in its investigation of the persistent outbreaks of famine (PWD 1903). In response to a question about the unsatisfactory condition of famine relief works, the acting chief engineer of irrigation (Col. Smart) made the recommendation that "a Revenue and a Public Works officer should be placed on special duty to jointly revise the programmes and, in consultation with the local officers, to ascertain local wants in the way of village tanks and wells and other works" (143). Well into the interrogation, a committee member returned to this recommendation and asked, "Why cannot the famine relief programmes in the district be jointly revised by a local district and Public Works officer?" (148). The response that follows illustrates the institutional dysfunctions that undergirded the colonial state:

> [Col. Smart]. "You have to find out new works; how is that possible?
> Q. Surely the local Revenue and Public Works officers can find them out; then what is the good of putting on special officers?
> [Col. Smart]. "They have too much to do. The Collector is immersed in office work and never goes out in company with the Executive Engineer. That is one of the wants of the system." (149)

This brief exchange illustrates some of the central institutional limitations of the state's bureaucratic approach to water, agriculture, and the outbreak of famine. The centralized colonial state, depicted here in the form of the Irrigation Commission, remained far from the everyday institutional quandaries and quarrels between the institutions tasked with managing the two key state objectives—the continuing drive to preserve agricultural productivity through irrigation and the extractive objective of revenue collection.

The chief engineer's response to the questioning embodies the institutional framework that had become embedded in the colonial water bureaucracy. From the perspective of the PWD, the solution to the famine lay in its technical expertise. On the one hand, this meant that "new works" were the automatic and self-evident response. One the other hand, this meant addressing the revenue office's dissociation from the practical technical problems of irrigation works. The chief engineer's negative characterization of the "office work" of colonial administrative officials such as the collector reflects the significance of technical expertise and fieldwork that the irrigation bureaucracy claimed as a basis for its authority and superiority. The colonial administration of the PWD in India illustrates a set of underlying tensions between the practical and political imperatives of the colonial administrative apparatus and an expanding field of professional knowledge. Such tensions are often overlooked by intellectual frameworks that emphasize the ways in which professional expertise became a crucial foundation for British colonial administration (Mitchell 2002). The postcolonial narrative on the nexus between colonial state power and professional expertise often appears as a natural or inevitable trajectory for state power. What is missed in this story is the active institutional lobbying of bureaucratic organizations to develop and preserve this mode of state power.

Professionalism, Expertise, and the Institutional Power of the Public Works Department

As is customary with bureaucracies, one of the PWD's central objectives was to consolidate its institutional power through its own organizational strength. The Public Works Department's deployment of languages of professionalism and expertise were not simply designed to expand the reach of the colonial state over Indian society but to expand the organization's power within the institutional structure of the colonial state. Senior bureaucrats in the department engaged in persistent lobbying for an expansion of a professionalized staff and a corresponding expansion of educational training for prospective recruits. In the initial years, the organization's staff drew on military engineers, with only a few civil engineers sent from England, and early works were focused on military projects, such as the provision of water supplies for British troop barracks (PWD 1868, 1). Once the colonial state had, in the phrasing of Viceroy John Lawrence, "finally accepted" the centrality of irrigation works for its political and economic interests, the PWD began to gain strength as an autonomous organization of civil engineers with its own professional bureaucratic interests (20). The PWD began to see a steady and significant growth in its staff (see table 1.1). By 1896, the colonial government (secretary of state) had authorized an expansion of the permanent engineering staff from 730 to 838 (PWD 1898, 1). While railways remained dominant within the PWD, given that the state had accepted irrigation as an "obligation placed upon the Government" (27), the irrigation bureaucracy became a central part of the PWD's administrative apparatus. As part of this expansion, a separate branch of the PWD with a special focus on irrigation administration was set up in each of the provinces under British rule (20).

An analysis of this expanding institutional power of the Public Works Department cannot adequately be captured by a self-evident explanation of the centralized colonial state's acceptance of the power of professional expertise. Archival records show that local administrative structures had

TABLE 1.1. PWD engineering staff in colonial India

Year	PWD engineering staff
1864	555
1868	783
1889	730
1896	838

SOURCE: Tabulated from *Summary of the Principal Measures Carried out by the Public Works Department* (Calcutta: Public Works Department Press, 1868 and 1898).

to persistently press both for an increase in the engineering staff and for employment compensation. If the PWD had made its case for public works as a critical component of state power, it did not inevitably receive resources for the successful implementation of such works. Civil engineers, for instance, were, in the early years of the department's history, paid less than were military engineers as well as military officers without a claim to professional expertise. Members of the PWD wrote in a petition to the viceroy that they, "although civil engineers professionally trained, and in a department where a training in civil engineering is above all things requisite, are, as a rule, whilst performing exactly similar duties, and holding positions identical in responsibility, allowed less pay than is given to military officers in the department, who (with the exception of military engineers) have not had the advantage of such training" (PWD 1869, 8).

Claims for monetary compensation were also accompanied by cultural grievances about the ritual practices of institutional status, with engineers complaining that "on any public occasion, when officers of the several services are assembled together, the civil engineer has no recognized status whatever, whilst the military officer, his junior in the department, takes precedence according to his military rank with the officers of the civil, medical and ecclesiastical services" (10). By the beginning of the twentieth century, records of the Madras Presidency administration still noted that the PWD "is undermanned and requires the immediate introduction of capable officers of all grades. Considerably higher salaries will have to be offered to secure the services of officers of the desired qualifications" (MP 1907, iv). Expertise, in practice, was not uniformly valued by the colonial state, nor was the investment in the administrative cadres representing the new values of colonial modernity easily forthcoming from the upper echelons of the colonial state.

The PWD's efforts to expand its organization through its claims of expertise also rested on a set of racialized colonial discourses. The professional cadres' claims for resources of newly emerging fields such as civil engineering were based on a logic structured by racialized distinctions between English and "native" expertise. Consider, for instance, the civil engineers' claims for increased pay. British civil engineers petitioned against employment classifications that placed them in "the general uncovenanted service, with clerks and other subordinates," arguing that "it has been fully established that the engineer branch must consist chiefly of Europeans, and a distinction which subjects European civil engineers in the Department of Public Works to rules originally framed to meet the requirements of a subordinate native service ... [was] a great hardship" (10). Professional expertise in this conception was not a self-evident field but the product of racialized institutional norms. While the prevalence of racialized discourses at the heart of the colonial state is not surprising, the point of significance lies with the ways in which technical and professional expertise was reconstructed through racialized distinctions. This distinction was not simply about contrasting colonial modern engineering expertise with local indigenous practices. Rather, expertise was structured through organizational practices that constructed Indian engineering and practical expertise as inferior.

Consider an early exposition on the field of irrigation-related civil engineering that was produced through the Public Works Department. As one report put it, "It is, I believe, too common an idea in England, that the natives of India are without an engineering history, that there are no works extant of their engineering skill, and that they owe to us all that they possess in that department: such is not the case. India *has* an engineering history; not written in splendid palaces and lofty structures, yet still marked by works whose usefulness may vie with the works of any other nation—works on which her life depends" (Tyrrell 1873, 1).

Writing specifically about irrigation works in South India, the report admitted that "the original idea of the annicut or bank or wall across a river is nevertheless a native idea; and so are tanks or the storage of water.... As the natives did, we did. We made annicuts on the low plains and took off channels for irrigation" (19). However, the report quickly went on to note that "the ancient engineering works of India in the south are, with the exception of tanks, neither very numerous nor well executed" (2). The spectacle and power of large works of irrigation were both central to the economic vision and the theater of state power embodied in such works in contrast to the colonial view of Indian engineering as "petty works" (Cotton 1854, 265). What is of interest in this colonial narrative is the momentary recognition of an existing history of technical engineering expertise. The report would go on to make a vigorous case for the professionalization of the field of engineering through organizational reforms within the PWD and the expansion of education aimed at "the development and advancement of the intellectual and practical knowledge of the natives" (15). This expansion would indeed take place and form the basis of a modern professionalized model of engineering and science.

The weight of the racialized logic of the colonial narrative concealed this existence of an engineering history prior to British rule.[3] Consider the passing reference to the system of tanks in South India. The elaborate system of tanks that has been the central foundation for irrigation was constructed by the end of the eighteenth century (Mosse 2003). To write this technical history out of a discussion of expertise and colonial modernity is to produce an artificial dichotomy between local customary village practices on the one hand and centralized state practices on the other. Images of the traditional autonomous village were in fact themselves a product of colonial ideologies (Ludden 1992; Mosse 1999). The tank system in the region was always a key arena for the exercise of state power. The imposition of colonial rule, as Mosse has put it, had "more to do with changing systems of state than the erosion of village tradition" (2003, 11). In this context, the rise of the disciplinary formation of engineering expertise was also not a novel development. What was distinctive was the way in which the institutionalization of the field sought to erase existing histories of engineering and relocate the practical and intellectual knowledge of civil engineering within the institutions of the colonial state.[4] The displacement of Indian staff by Europeans in the PWD was not a product of the technical field of engineering but the result of specific organizational policies that were shaped by racialized understandings of labor and skill. The disciplinary formation of the field of engineering in India (through the educational institutions that were set up by the colonial state) was a product of this set of bureaucratic practices designed to both protect the interests of European civilian staff within the PWD and expand the power of this bureaucratic wing within the broader apparatus of the colonial state. Expertise and knowledge formation were the means through which bureaucratic organizations such as the PWD could consolidate and expand their institutional power.

The persistent institutional claims and demands of bureaucrats within the irrigation and water-related wing of the PWD would ultimately transform this agency into an essential component of colonial state power. The local bureaucratic power and lobbying of the organization soon became a key framework of the centralized colonial state. In this process, state power was shaped by a complex historical relationship between the formal centralized colonial state, the centralization of state power through local bureaucratic practice, and invocations and representations of the welfare of the public.

The Water Bureaucracy, the Public Interest, and the Tributaries of State Power

The colonial water bureaucracy was founded on the principle of developing a centralized mode of state control. The first report on the newly emerging system of public works of irrigation transparently stated that the administrative organization over water resources and infrastructure in the Madras Presidency was designed "to place the whole course of every river, as far as practicable under one control" (GC 1856, 5). This centralized, technocratic vision of river and water control that the PWD used to consolidate its institutional power within the colonial state provided a firm foundation for the organization to serve as an expansive arm of colonial state power. However, this centralization happened through the local bureaucratic structures of the department. The irrigation bureaucracy of the PWD soon succeeded in steadily expanding its authority and power within the state apparatus. On the ground, the organization gradually increased its effective authority through the control over both water and land. The distribution of water for large works of irrigation, for instance, was placed under its control (GM 1858, 33). While the sanction of works permitted collectors to undertake some repairs, repairs of irrigation works could not take place without the authority of the PWD, thus continuing to expand the PWD's power in relation to the more decentralized administrative apparatus of the Revenue Department.[5]

At the heart of this mode of state power was a discursive frame that increasingly conflated the interests of the state with the interests of public welfare in India. The "publicness" of "public works" was a central rhetorical device that allowed for the state's self-interested control and appropriation of water resources in the name of society and the public good. This modality of state power transformed public infrastructure into a material-discursive realm that melded together the realms of state and society in colonial India. Water infrastructure was the central site for a deep and invasive method for the colonial institution of the PWD to permeate Indian society while claiming to serve the public interest.

Such invocations of the public interest also provoked moments when the state had to account for its rhetorical claims. The deleterious effects of famine compelled the colonial state to take into account the ways in which irrigation works were addressing the subsistence needs of the population it ruled over. The Famine Commissions, for instance, called for a shift from irrigation works focused on productivity to a focus on famine and the subsistence needs of the people.[6] This led to the development of a new category of "protective works" that were distinct from the extractive "productive works" of irrigation designed to serve the state's revenue-collecting objectives. However, a closer look at the underlying framing of such protective works illustrates that they remained within a monetized framework linked to the state's primary objective of revenue extraction. The category of "protective works" was specifically designated to focus on famine protection by reducing the threshold for revenue returns (protective works required a 3 percent rate of return; HD 1905, 6). As a report on the protective works in the Madras Presidency noted in the case of the "Kistna" (Krishna) Delta system, "It is, I believe, universally admitted that the duty of water, in this district, as elsewhere, is much lower than it should be and that it is capable of being considerably increased. The efficiency of the regulating and distributory works has much to do with the economy of water, but still more does it depend on the *personnel* of the controlling staff" (MP 1902, 6).

Yet the investigation for the state's Irrigation Commission during the same time period yielded at least one case in which an official responded that water charges were discouraging villagers from growing a second rice crop (PWD 1903, 155). Thus, despite the colonial government's claims that its irrigation policies were designed to serve both the revenue interests of the state and the welfare of the public, such microinstances reveal the ways in which the monetization of water would overshadow the claims of welfare, in this case by discouraging crops that could aid in the state's goals of famine protection. The focus of the administration's criticism of the "personnel" in increasing water charges further illustrates that despite the centralized state's professed concerns regarding public welfare expressed in sites such as the Famine Commissions, in practice, it expected the "personnel" of the PWD to execute the state's primary objective of revenue collection.

The monetization of water was in fact a central foundation of the PWD's approach to the governance of water. The first attempts at monetizing water in the Indian context were set in motion by the PWD. As early as the mid-nineteenth century, reports of the PWD began to conceptualize the "calculation of the value of water" through the currency of "money valuation" (Cotton 1854, 170). Writing about the comparable availability and value of land and water, Cotton argued for a colonial state policy in which, given the limited availability of land at the time, "The *water* therefore should be sold, and little more than a nominal rent charged on the land" (185). While changing land markets under colonial rule altered this initial assertion about land rents, what is of significance in these early writings is the establishment of the early foundations of the commodification of water. Cotton's report for the PWD is interspersed with elaborate calculations that seek to measure and prove the monetary value of water. In this vision, the objective was no less than the grand assertion that in doing so "the total amount of treasure in the country, in the shape of water, may thus be calculated" (213). This monetized conception of water factored directly into the calculations that the East India Irrigation and Canal Company made in their assessment of the profitability of investing in water-related infrastructure. The company listed the main sources of profit as "The Sale of Water to the Government for the Irrigation of Land" (EIICC 1860, 10). As the company's prospectus noted, "Indeed, it is difficult to conceive how the sale of an article so absolutely essential to life and progress as water is everywhere, but more particularly in India, can be otherwise than productive of large profits" (10).

The emerging conception of water as a form of capitalized resource remained foundational for the colonial state's conception of the "publicness" of water and water-related infrastructure. The institutional framework of the PWD, in effect, solidified the conflation of the "publicness" of water infrastructure with the interests of the state. For instance, from its inception, the department asserted its authority over water resources. The state was opposed to the digging of private wells (Cotton 1854, 264) and the construction of private tanks and sought to place constraints on such endeavors by ordering personnel to ensure that "the position of it shall not compromise existing rights, either of Government or of private persons" (Grant 1857, 131).

The inextricable relationship between water and land, underlying this example, provided numerous channels for the PWD to exert and expand its authority. Colonial records show that the department was able to exercise its power in ways that crossed the boundaries of its own institutional purview. As early as 1855, internal reviews showed that the state had concerns about cases in which "channels have been taken [by the PWD] through Zamindary lands, without the permission of the owner, and without any compensation" (177). While such internal reviews ultimately did little to check the practical power of the department, they provide important examples of the PWD's interventionist mode of state power. This centralized form of state power was exercised through local institutional practices of bureaucrats.

The expanding bureaucratic tributaries of the PWD's power also formed the basis for the formal legal architecture of the colonial state's governance over water. A series of irrigation acts in the nineteenth century codified the colonial state's project of harnessing water for production (Cullet 2009).[7] The state control of water resources that was being consolidated through the construction of public works and the collection of water charges and revenue later culminated in official legal control through the Madhya Pradesh Irrigation Act (1931). This act formalized the state control that had become entrenched through the practical authority of the PWD. The underlying control over land through irrigation infrastructure that was emerging through the PWD also led to the beginning of a regime of property rights that was centered on the control of water resources (Cullet 2009).[8]

By the time of such formal codification of the legal rights of the colonial state over water resources, in the early decades of the twentieth century, the irrigation wing of the PWD had been firmly established as a central arm of the state apparatus. This power of the PWD subsequently expanded beyond the formal territorial boundaries of the colonial state in the Indian subcontinent. This is well illustrated in the role of the PWD in water disputes and negotiations between the colonial government of the Madras Presidency and the neighboring princely states in South India. Tensions over water sharing grew by the 1930s, as increased water infrastructure began to intensify competing demands on water resources of the Krishna Delta and the Cauvery River resources. PWD officials became key interlocutors of the colonial state as officials of the civil service turned to them for guidance on such matters (PWDI 1938). A key area of contention was the construction of infrastructure that made claims on shared water resources between the colonial state of Madras and the princely states of Mysore and Hyderabad.

Consider, for instance, the growing tensions over water resources and infrastructure between Madras and Mysore. Such tensions began with the change in Mysore's status with the Rendition Act of 1881, which reversed British annexation of the princely state. During the period of British rule over Mysore, the "Chief Engineer of Mysore submitted a comprehensive scheme of developing irrigation from the waters of all of the rivers of the state" (SB 1935, 3). The engineer attested that there would be no damage to the interests of the Madras Presidency and proposed to construct a large reservoir at Lakkavalli for protection against drought (3). However, after the rendition, as one PWD letter noted, "The adjudication of rights over waters and the protection of different interests assumed great importance" (SB 1935, 3). In 1892, an agreement was signed between the princely state of Mysore and the colonial government of Madras that sought to preserve British colonial interests (SB 1935, 3). According to the terms of this agreement, Mysore agreed to certain restrictions that specified that the princely state would not construct new works of irrigation or reservoirs without the consent of the Madras government on fifteen rivers as well as forty-five streams and drainage areas (3). In return, the colonial government agreed to the construction of the Lakkavalli Reservoir and promised to consent to new irrigation works that did not impact its interests. Given the large number of rivers flowing from Mysore into Madras, this agreement became a crucial means for the Madras government to protect its interests. As the PWD correspondence noted, "By these provisions Mysore gave up her claim to use water within her territory as she pleased and Madras acquired a very valuable right, which she has frequently exercised to control irrigation schemes in Mysore territory" (4). The PWD was a crucial actor in this expansion of colonial state power over water resources, as the department provided the detailed calculations and technical arguments that colonial administrators used as the evidentiary basis for pressing its interests. In the process, the collection of data and technical expertise allowed the PWD to further consolidate its institutional power within the state apparatus.

The dominant position of the colonial state over the Mysore princely state encoded this relationship of power within this agreement. As the centrality of public works of irrigation for the political and economic interests of the colonial state grew, such water-based infrastructure became a site for the continued pursuit of state interests. By 1905, the Madras government had developed plans for a new reservoir that put it in competition with Mysore's Lakkavali Reservoir. Citing the public welfare of "several million of his Majesty's subjects [who would] be imperiled through famine and privation," the government reversed its agreement on Mysore's reservoir (5). The result was a prolonged negotiation between the two states that exemplified an expanding set of conflicts between the Madras government and its neighboring princely states. In the case of the princely state of Hyderabad, similar conflicts over irrigation infrastructure began to arise. As a letter from the Political Department of the Hyderabad state noted, the "equitable distribution" of shared water resources with Madras was a significant source of contention, with Hyderabad contending that the Madras state had expanded irrigation in violation of agreements between the two states.[9] The Hyderabad government protested "the steady expansion of irrigation in Madras territory without regard to the summer flow available in the river" and argued that "Hyderabad holds that rights do not merely arise out of *de facto* appropriation.... The cardinal rule underlying the relations of one state to another is equality of right."[10]

The intensification of state claims over water resources set in motion an array of interstate negotiations, arbitration processes, and tribunals between the colonial presidencies and princely states in southern India. The colonial state was prohibited from simply imposing its will on the princely states despite its political and economic dominance (Mollinga 2003, 103). The British state instead tried to assert its dominance by attempting to control the terms of the arbitration. However, this also produced tensions within the colonial state, as the Madras and Bombay Presidencies each sought to protect their own interests. By the 1930s, the Madras state sought a single set of arbitration proceedings that addressed all interstate disputes that included the Bombay Presidency, Hyderabad, and Mysore (PWDI 1938). However, both the Bombay PWD and the Hyderabad Presidency sought to disentangle themselves from the protracted disputes between Madras and Mysore. The process of negotiation produced a certain level of compromise between the various state entities. While a conference between the four states took place, it was accompanied by two sets of bilateral negotiations between Madras and Hyderabad and Mysore.[11] The negotiations ultimately produced an agreement in which the Madras state made some concessions in return for moving ahead with its desired projects, such as the harnessing of the Tungabhadra waters for irrigation that it had been pursuing for decades. Their concessions included the allotment of waters from the Lakkavalli Reservoir and what the Madras state considered "a very substantial concession" to Madras in the form of the reduction of a royalty payment for electric generation from Cauvery River waters (at Sivasamudram).[12]

While the Madras Presidency's political negotiations through formal arbitration reveal the complexities of interstate relations despite the dominance of the colonial state, they also point to the deeper ways in which the construction of public works nevertheless provided a crucial means for the exercise of colonial state power. Such negotiations took place in the context of the decades of the PWD's infrastructural work that had already placed Madras in a position of dominance in the use and control of water resources. The centrality of the PWD as a means of colonial state power was further intensified in the course of such interstate negotiations. At one level, engineering assessments formed a foundational component of the evidence that the Madras state used in its political negotiations. At a deeper level, engineering experts also laid out the political principles of water sharing that favored the interests of the colonial administration. Consider, for instance, one of the central principles of water sharing that the chief engineer delineated: "The quantity of water to which a state is entitled is limited to the quantity she needs for concrete schemes which she proposes to put in hand but unutilized waters should not be allotted to any particular state but should be kept free to meet demands for the concrete schemes of a Sister State. In other words, the available waters should be used to the greatest good of the greatest number" (PWDH 1937, 36).

This delineation of water-sharing principles illustrates the architecture of the colonial state's control of water resources. By foregrounding the links between the right to water resources and "concrete schemes," the chief engineer reinforced the specific institutional interests of the PWD as the chief organization in charge of such schemes. In doing so, the engineer in turn reinforced the broader state interests of the presidency. The early and expanding economic dominance of the Madras Presidency in advancing public works of irrigation gave the state an effective advantage on the ground and strengthened its authority. Infrastructural state power was once again put forth in the name of public welfare by ostensibly serving "the greatest good of the greatest number." In the process, the fusion of the institutional power of the PWD's water bureaucracy, the interests of the colonial state apparatus, and the state-defined conception of the "publicness" of public works was once again cemented.[13]

While the colonial state continually sought to use water infrastructure as the material-symbolic embodiment of its claims of serving the public interest, the extractive nature of its governance over water resources also provoked civil resistance within the Madras Presidency. There were growing civil protests against rising taxes that the state levied for lands benefiting from irrigation works (Stoddart 2011). If the new land/water nexus was a basis for expanding state power, it also became the source of new forms of political mobilization and associational organizations in the nineteenth century. Such forms of protests ranged from petitions against higher taxes on land to proposed increases in water rates (Stoddart 2011, 10) to the "widespread relinquishing of irrigation water in many delta villages" (Stoddart 2011, 12). By the early decades of the twentieth century, such localized protests eventually fed into more systemic forms of nationalist resistance. In the Guntur District of the Madras Presidency, "Local Congress leaders linked yet another revenue no-tax-payment campaign with Gandhi's mission, the district becoming both the presidency's 'hot spot' and one of the leading national sites, the citizenry refusing to pay land and water taxes. Local grievances dovetailed perfectly with national aspirations. The importance for national politics of the preceding land and water campaigns was obvious" (Stoddart 2011, 24). The reach of the colonial water bureaucracy thus produced a corresponding network of resistances that became nationalized in the twentieth century. The public interests of such infrastructural works became the site for a continual space of contestation.

By the mid-twentieth century, India's newly independent nation-state would inherit this contested "public" nature of water infrastructure and governance. On the one hand, the institutional structures of the water bureaucracy—ranging from its organizational practices and traditions to its regime of knowledge and expertise to the formalized legal frameworks—became a part of India's new postcolonial state. On the other hand, the contestation over the land-water nexus that fed into the nationalist movement meant that the new nation-state was also shaped by a competing set of social and economic priorities and visions. The statist model of public works of irrigation, the reach of bureaucratic authority, and the contested welfare of the public was fully solidified by the mid-1940s. It is in the context of this weighty historical institutional context that India's developmental state's approach to water resources was forged.

India's Developmental State and the Governance of Water

In the first decades of independence, India's regime of water governance was shaped by a complex configuration of the underlying political and economic structures that had been consolidated under centuries of colonialism and the new priorities and policies of the postcolonial state. The historical formation of India's postcolonial water state can be understood through a focus on three central areas—the nature of the federal regime of governance over water resources, the policies and priorities of the developmental state, and the practices of local water bureaucracies. While it is now commonplace to depict the early decades of the political economy of the Indian state through the image of a highly centralized, autonomous state, this image is unsettled by the state's mode of governance over water. What emerges instead is a multifarious set of characteristics that combines the command-oriented developmental state with a weaker and undeveloped set of federal structures designed to manage water resources. This antipodal nature of the state was in turn enmeshed in the political-economic structures produced by colonial rule. The centralizing nature of the newly independent command-oriented state was shaped by the complex relationship between the power of the central government and the centralizing nature of state authority within local state governments. Centralized state power associated with India's planned developmental state was marked by significant national regulatory institutional gaps. Both the mode of centralized planning and these national institutional gaps in turn intensified local forms of centralization within state governments.

The federal structure for governance of water resources was marked by a set of distinctive features that make it a rich case for an understanding of these dynamics of state authority. The state's approach to water governance contained within it a number of ambiguous and contradictory facets that were distinct from the kind of centralized federalism (Sharma and Swenden 2017) that characterized the Indian state in other arenas of politics and the economy. The formal constitutional framework lists water as a subject that is in the purview of the authority of both states and the central government (MWR 2018). While state authority over water has generally resided with local state governments, authority under the Union List allowed the state to specifically carve out an exception for the central government to exert its authority over interstate rivers when such authority serves the "public interest."[14] This exception was codified into the Inter-state River Water Disputes Act, 1956, in accordance with Article 262 of the Constitution. The resulting federalized structure for water governance contained an inbuilt institutional weakness in the authority of the central government (Iyer 2002). First, the role of the central government in overseeing interstate resources was focused on the management of conflicts rather than on a more productive role of facilitating interstate cooperation (D'Souza 2009). In recognition of this institutional failure, the Ministry of Water Resources eventually recommended a replacement of the River Board mechanism, noting that the "Central Government can constitute a River Board under the provision of the River Boards Act, 1956 with the concurrence of the State Governments. The Central Government has however not been able to constitute any River Board under this Act so far. The role of the River Boards as envisaged in the said Act is only advisory in nature. The National Commission for Integrated Water Resources Development Plan has recommended the enactment of a new Act called the 'Integrated and Participatory Management Act' in place of existing River Boards Act, 1956" (MWR 2002, 65).

The Indian state's institutional architecture was historically oriented toward the mediation and resolution of disputes once they had arisen; there was no policy framework that proactively promoted models of planning and development that could build and strengthen interstate cooperation over water resources (D'Souza 2009, 89). The consequences of this institutional vacuum at the level of the central government were that state governments were left to harness local water resources until water scarcity provoked conflicts with competing state governments. The central government was then compelled to intervene once the conflicts had already accelerated and in many cases been politicized. This pattern has continued in the postcolonial period.

This distinctive federal structure that shaped governance of water provides rich terrain for a rethinking of how centralized state power has been exercised in contemporary India. Given that the subject of water was placed in a decentralized framework since the early years of independence, an analysis of the dynamics of water governance allows for an analysis of the centralization of state power that is not simply conflated with the authority of the central government. In the case of water, processes of centralization have, in practice, been shaped by the power of both the central government and bureaucratic organizations within local state governments. Consider the realm of the regional governance of water. The institutional gaps at the national level exacerbated the unregulated appropriation of shared water resources that in turn intensified and produced significant interstate water disputes in the first decades of independence. However, while such disputes have laid bare the institutional inadequacies of the central government's regulatory functions, they also point to the more subtle forms of centralized state control over water resources at the local level. Local state governments and their water bureaucracies have increasingly sought to intensify control of water resources, both as they have competed with neighboring states for resources and as they have served as the central arbiters of the distribution of resources between competing demands for water.

Such nuanced and often less visible patterns of local state authority were accompanied by more familiar forms of centralized state control over water. The planned economy of India's postindependence state rested on the intensive extraction of water resources in pursuit of India's developmental goals. India's developmental state in the twentieth century approached water resources through a purely instrumentalist policy framework. Water needed to be harnessed in pursuit of the state's goals of achieving food security and accelerating industrialization. Large dams were not just the symbols of the Nehruvian modernist vision but the material infrastructure that embodied the state's approach to water, irrigation, and agricultural development.[15] The combination of the absence of a national institutional framework for the management of water and the new developmental objectives of the postcolonial state meant that water management was driven by the development and implementation of public works (Raina 2015, 339). However, as with the case of interstate water disputes, while the state was the central actor in shaping the uses of water through its centralized investment and development policies, this centralized authority was not exercised in conjunction with adequate national regulatory frameworks. Such incapacities were deepened by a fragmentation of institutional governance, as the regulation of surface water and groundwater were under the purview of two separate institutions, the CWC (Central Water Commission) and the CGWB (Central Ground Water Board). In addition, in the 1970s, there was a rapid expansion of inefficient groundwater irrigation systems (Dubash 2002; Frankel 2015). The result was the creation of deep-seated problems that have become hallmarks of the challenges that continue to shape the governance of water in contemporary India.

Consider, for instance, the ways in which the state entrenched an extractive model of groundwater used for irrigation. From 1950 to 1997, "nearly 4/5ths of public expenditure was for irrigation," but while 70 percent of the state's expenditure "was for surface irrigation purposes, groundwater provided the largest share of irrigation water" (Dubash 2002, 4). The expansion of groundwater extraction started in 1965, and by the 1980s, the rate of expansion of groundwater irrigation outpaced that of surface water (Biswas and Hartley 2017). This model of intensified groundwater extraction was also supported by international development agencies, such as the United Nations Development Programme. In the case of Tamil Nadu, the number of diesel and electrical pump sets used to pump groundwater increased from 527,530 in 1970 to 1,719,817 in 2001 (GTN 2003, 131). Such processes have had significant implications for agrarian communities (Dubash 2002; Ghuman and Sharma 2018). This extractive approach to groundwater has continued to shape the governance of water in Tamil Nadu in the context of urbanization, with serious implications for the effective long-term management of the state's resources in times of water scarcity.

The absence of effective centralized regulatory frameworks also characterized other areas in the first decades of developmental planning. While centralized structures designed to govern have been in place since the formation of the Central Waterways, Irrigation and Navigation Commission (CWINC) in 1945 (Shah 2016, 69), a more focused approach to water did not evolve until the 1970s. The Central Water Commission, which evolved out of earlier structures, was formed in 1974, and as the Mihir Shah report has noted, while the Central Ground Water Board was established in 1971, "it was only in the latter part of the 1980s that groundwater assessment began to take shape in CGWB's thinking" (2016b, 88). Broader water laws evolved later or in response to the effects of policies of development rather than as part of an integral part of the planned economy. For instance, national rural water drinking guidelines were not established until the 1970s, and the legislation for the prevention of water pollution was not passed until 1974. Governmental policies across the political spectrum ignored water quality concerns and increasingly focused on the needs of industry over farmers despite the enactment of the 1974 Water Prevention and Control of Pollution Act (Saravanan and Appasamy 1999, 177). Adequate central regulatory institutional frameworks were not a major feature of the kind of state centralization that was taking root in India's planned economy.

Delayed and inadequate regulatory frameworks in the postindependence period underlined the significance of the colonial legacies of water governance. The core legal framework of water governance after independence continued to largely draw from colonial-era laws. The result of this approach was that the new goals of the developmental state were overlaid onto the longstanding legal and institutional structures that had been developed by the colonial state. In the colonial era, the state had established its authority over water resources. In the case of the Madras Presidency, the Madras High Court had pronounced that the government had a sovereign rather than a proprietary right over the supply and distribution of irrigation water (Upadhyay 2009, 138). Later legal decisions by courts such as the Madras High Court and the Supreme Court in independent India would uphold the authority of the state to determine the regulation and use of water resources (Vaidyanathan and Jairaj 2009). As in the colonial period, in practice, the embodiment of this confirmation of the sovereign right of the state over water resources was the local water bureaucracy.

Such historical continuities from the colonial era also shaped local institutional practices. In the case of the Madras Presidency, departmental organizations such as the Revenue Department and the PWD retained their administrative power within the new state of Tamil Nadu in independent India. Such institutional continuity brought with it the deep-seated institutional patterns that had been established during the colonial period. The internal institutional fissures and fractures between competing departments, such as the Revenue and Public Works Departments, remained a consistent feature of postcolonial state administration. At one level, the detailed paper-intensive system of reporting that emerged in the navigation of these interinstitutional rivalries became a distinctive characteristic of the administrative state and a century later became the foundation for the infamous image of India's sluggish postcolonial bureaucracy. At a deeper level, the state's institutional fragmentation deepened the deterioration of Tamil Nadu's network of tanks (Mosse 2003, 46). This deterioration of the tank system was further compounded by the state's adoption of the central government's developmental model that promoted groundwater extraction and larger public works, such as the extension of canal irrigation.

This reconstitution of colonial institutions meant that the powerful local bureaucratic organizations retained their authority over water resources but now executed this authority in conjunction with the developmental goals of the postcolonial state. The PWD in Tamil Nadu served as the central bureaucratic arm for the implementation of state developmental policies. According to government regulations, "The officers of the Public Works Department exercise complete control over the distribution of water in the larger works of irrigation" (GM 1958, 123). This realm of authority, as in the colonial period, also gave the PWD authority over land related to water-based infrastructure. Such authority encompassed a wide range of activities, including the ability to negotiate with landowners, to initiate land acquisition proceedings through the 1894 Land Acquisition Act, to lease lands for infrastructural projects, and to make grants for occupation by both private individuals and companies (GTN 1986, 64–66).

This purview of the PWD's authority intensified the organization's institutional investment in large works of irrigation—an orientation already circumscribed by the disciplinary practices of civil engineering that had become the sole source of training for employees of the water bureaucracy (Mosse 1999, 2003). The PWD, for instance, continued to use its authority over water distribution to regulate cropping patterns (Mollinga 2003, 63; Wade 1982, 299). Meanwhile, in the absence of policy guidelines from either the state or the central government on how to manage competing water demands for irrigation, drinking water, and industrial uses, such policy decisions were in effect practically made through the PWD's decision-making on specific infrastructural works. For example, the PWD sought to mediate conflicts between the irrigation needs of farmers and the urban drinking water and supply needs of the municipality of Coimbatore and took over the control of water infrastructure in the process. In the face of conflicts over the municipality's growing opposition to supplying irrigation water to farmers through a tunnel of the Siruvani Dam in 1951, "the government realized that if the PWD took over the maintenance of the tunnel, difficulties in diverting water for irrigation purposes could be solved.... [The] District Collector discussed the surplus water diversion at the district board meeting and the board had approved the inclusion of the scheme in the Second Five-Year Plan" (Saravanan and Appasamy 1999, 180).

TABLE 1.2. Tamil Nadu PWD irrigation budget, 1967–1973

Year	Expenditure on irrigation (original works) (rupees—millions)	Expenditure on irrigation (maintenance) (rupees—millions)	Total outlay (rupees—millions)	Percentage of PWD budget spent on irrigation (%)
1967–68	69.8	33.8	140.5	73.7
1968–69	64.8	35.1	137.2	72.8
1969–70	70.1	35.3	155.5	67.8
1970–71	82.5	37.6	196.6	61.1
1971–72	85.8	40.6	213.5	59.2
1972–73	109.9	35.1	225.3	64.3

SOURCE: Administrative Reforms Commission, *A Report on the Public Works Administration*, vol. 1 (Madras: Government of Tamil Nadu 1973).

This microinstance illustrates the centrality of local bureaucratic assertion in the management and control of water resources. It is this local form of state authority and the objectives of the PWD that are incorporated in the central government's five-year plan. In contrast to conventional understandings of the top-down nature of the planning process in the early decades of independence, this example shows how the local water bureaucracy was able to consolidate its authority through the centralized planning process. In effect, the dynamics of centralizing state authority flow from the local level to the central government. This intensification of local bureaucratic control was also manifested in broader patterns of funding, as central planning also created a steady and substantial increase in the PWD's irrigation budget (see table 1.2). Thus, from 1967 to 1973 the PWD's budget increased from Rs. 69.8 million to 109.9 million.

Such patterns and practices illuminate the contradictory nature of state authority that emerged in the era of the twentieth-century developmental state. The strong role of the central government in India's planned economy was intertwined with significant regulatory gaps that in turn intensified local bureaucratic control over water resources. In other words, the government's approach to water was shaped by a strong centralized approach and weak centralized regulatory frameworks. Centralization meant a command approach rather than strong regulation (L. Rudolph and S. Rudolph 1987). This distinction between the need for a strong central government regulatory framework and the need for centralized state authority is critical in unsettling the conflation between centralization and the spatial scale of authority by the central government. Meanwhile, local water bureaucracies became the primary vehicles both for implementing the centralized form of state planning of the developmental state and for stepping in to fill the regulatory gaps of the central government. Local water bureaucratic institutions, such as the PWD, played a central role in executing developmental goals in ways that both maintained and extended their power. State centralization was shaped by a paradoxical process in which inadequate national regulatory mechanisms (such as those that could balance different demands for water and those that could regulate pollution) allowed centralized state control to take hold through institutions within local state governments.[16]

This account of the water bureaucracy departs in significant ways from conventional accounts of the centralized federalism of twentieth-century postcolonial India. When it came to water governance, the commanding power of the state rested with local bureaucratic organizations, such as the PWD. This was reinforced by the weakness of centralized regulatory frameworks. The project of harnessing water was a means for the instrumentalist pursuit of economic goals. This instrumentalist approach, which transformed water into a vehicle for the state to achieve its development goals rather than a natural resource that required an autonomous and effective regulatory institutional framework for its preservation, meant that there was in fact little systematic national planning when it came to water policy, in contrast to the planned approach to the economy of the twentieth-century state. Indeed, India's first national water policy was not adopted by the Ministry of Water Resources until 1987.[17]

Such underlying patterns in the mode of state control over water were not limited to the model of large-scale public works of irrigation, which were at the heart of the postcolonial state's approach to irrigation and development, but more deeply embedded in the patterns and practices of the state's administrative apparatus. The organizational history of the Ministry of Water Resources itself reflects the fractured approach to water resources (MWR 2018). Water resources were primarily subsumed under irrigation, and the purview of irrigation was transferred between a series of governmental departments in the early decades of the postcolonial period. The subject of irrigation was first located in the Department of Works, Mines and Power then moved to the new Ministry of Natural Resources and Scientific Research in 1951, only to be recast into the Ministry of Irrigation and Power in 1952.[18] The growing political and economic significance of irrigation works led to yet another reorganization with the establishment of the Department of Irrigation in 1974 under a newly independent Ministry of Agriculture, which directed the command-oriented approach of the developmental state. As I have noted earlier, the CWC and the CGWB were also established during the 1970s. It was not until 1985 that the Department of Irrigation was reconstituted as the new Ministry of Water Resources (MWR 2018). However, by the mid-1980s, the overexploitation of groundwater resources in the service of the state's developmental aims, the intensification of states' competition over resources in water-scarce regions, and the historical legacy of local state control over water resources meant that the new national institutional regulatory framework was being layered over a dense set of political and economic structures that were already directing the management of water resources.

This overview of patterns of state authority illustrates that the nature of state centralization in postindependence India unsettles conventional center-state frameworks. It underlines the need to distinguish between regulatory frameworks of the central government and the centralized authority of the developmental state. Furthermore, the case of an organization such as the PWD illustrates the ways in which the planned developmental state produced complex forms of state power at the local level. While aspects of this state authority reflect conventional understandings of an interventionist central government, other dimensions point to weaknesses in the regulatory frameworks of the central government that expanded the space for centralization of control over water within local governments. The dynamics of centralization thus did not conform to spatial scales (where centralization corresponds to the largest scale of the central government), even within the heights of the command-oriented period of the developmental state. This significance of local state power is further evident in the more nuanced ways in which the local state permeated civil society in the twentieth century.

Developmental State Authority and Bureaucratic Class Formation

The embodiment of the state's sovereign control over water through local bureaucracies endowed organizations such as the PWD with the kind of local state power that expanded the space for the forms of practices of patronage and corruption that are now seen as endemic to the Indian state. Intricate layers of patronage and corruption were built into the construction and administration of irrigation works. Forms of illicit revenue derived from both farmers and contractors.[19] Such extractive payments were built into relationships between engineers of the PWD and contractors, as well as between contractors and politicians. As Robert Wade has illustrated, for farmers, "the use of a rotational delivery rule [from canal irrigation] can provide a pretext for a highly discretionary, predatory behavior by irrigation staff towards farmers," and in times of water scarcity the control of sluice gates could be used to protect from interference from downstream farmers or "vice versa for lower sluices: the officers can make sure the upstream sluices are not opened so that more is available for lower down" (1982, 299). These practices in turn deepened inequalities within rural areas, as well-off farmers or dominant rural groups with access to political power were able to exert pressure on local irrigation officers and staff (Vaidyanathan 1994, 42).

While the weight of such practices of patronage and corruption is real, the prevalence of bureaucratic corruption also stems from underlying historical practices of organizations, such as the PWD, that relied on local networks and clientelist relationships to complement the highly centralized structure of the colonial state. Political rhetoric and criticisms of bureaucratic corruption were produced through nationalist discourses and resistances (Gould 2011). Meanwhile, what would in the postcolonial period become constructed as the illicit revenue of corrupt bureaucracies was in fact built into the class formation of sections of the middle classes in twentieth-century India. Consider the following example from historian William Gould's research on Uttar Pradesh.

A retired PWD Executive Engineer:

> recalled how, by the 1960s, it had become quite commonplace, in arranging marriages, for the "extra percentages" derived from non-formal bureaucratic income contained within the bridegroom's salary to be taken into account, and that he had seen many examples of it. T.S.R. Subramanium, a retired IAS officer posted in UP, told the story of the wedding of a daughter of the executive engineer in the Public Works department in Ghazipur. His wife was closeted with the other women of the party and recounted that guests would come up to the mother of the bride and ask her about the salary of the prospective son-in-law who was also an assistant engineer in the same

department. The answer was invariably that his salary was rupees three hundred plus two percent. Subramanium concluded that "The system had allowed these [percentages] to become part of normal functioning." (2011 45)

The "extra" revenue of local bureaucratic officials was part of the historical social fabric of India's middle classes, which were primarily dependent on state employment in the twentieth century.

Such processes point to perhaps the least understood dimension of India's bureaucracy—a conception that focuses on the bureaucracy as the blurry ground that stands between "state" and "society." The bureaucracy in effect represented a central site for the state formation of India's middle classes. Both the late colonial and early postcolonial bureaucratic fields provided the foundational basis for middle-class formation. While moralistic views of corruption have often circulated as middle-class political discourses in both the nationalist and postindependence period, the extraction of revenue through patron-client relationships and practices of corruption was inextricably linked to the bureaucratic state's class formation of sections of India's middle classes.

This process of middle-class formation within the bureaucracy cannot, however, be adequately understood simply as a function of middle-class privilege and power within the developmentalist state. In the case of local bureaucracies, such as the PWD (in contrast to the elite bureaucrat forces, such as the IAS), middle-class employees often embodied contradictions within the state-class relationship of the developmental state. Take the case of the irrigation bureaucracy of the PWD. On the one hand, the scope for revenue extraction and the control over resources illustrates the wide scope in which local bureaucrats could wield state power, often in the service of their own private interests. On the other hand, the vast majority of employees also were in restrictive institutional environments that produced debilitating work environments. While the PWD in Tamil Nadu experienced a significant expansion of its workforce in the first decades of independence, employees often had little scope for upward mobility. For instance, by 1973, the PWD staff comprised 3 chief engineers, 19 superintending engineers, and 124 executive engineers at the higher rungs of the organization. Meanwhile, at the lower tier, the PWD had 665 assistant engineers and 2,050 section officers (including supervisors and junior engineers) (ARC 1973, 1–6;

data include both buildings and irrigation wings of the department). As an Administrative Reforms Commission report noted, most engineers stagnated at the same level with little scope for promotional opportunities (ARC 1973, 27). The promotion of engineers to the rank of superintending officer generally occurred near the age of retirement (18–19). Chief engineers, as the report went on to note, would consequently stay in their position for six months and thus have "no chance to provide the department with any dynamic or imaginative leadership" (19). Beyond these formal considerations, employees would have to pay for posts as well as for transfers in the late twentieth century, even as there was a "steady decline in the real value of engineers' salaries—by about half since 1965" (Wade 1982, 307). Illicit revenue in this context was a central class strategy that sections of the workforce used to preserve their middle-class status or gain access to avenues of upward mobility.

In addition to the financial and professional constraints of the formal terms of employment, middle-class bureaucrats often occupied a precarious position within the larger state structures of political patronage. While scholarly work on the politics of bureaucratic transfers has focused on the politicization of the elite tiers of the bureaucracy (such as the IAS), particularly in the post-Emergency period (L. Rudolph and S. Rudolph 1987), transfers in local bureaucracies were a means for engaging in extractive relationships *within* local state institutions. "Politicians and senior officers were able to obtain for themselves part of engineers' additional income by auctioning the transfer, and imposing additional demands as a condition of the successful bidder's not himself being transferred out before the normal term" (Wade 1982, 303–4). Processes of revenue extraction and patron-client relationships were thus not limited to the relationship between bureaucrats and the external public (whether citizens or contractors) but were part of the internal organizational framework of the bureaucracy.

Such local relationships were the material practices that forged the workings of the developmental state's water bureaucracy. Far removed from the image of an all-powerful centralized command state, the water bureaucracy was a product of a messier set of processes shaped by underlying historical patterns of the colonial state, the new institutional frameworks and policy agendas of the developmental state, and a variegated set of activities within the local bureaucracies that were enmeshed within these broader political and economic contexts. The water bureaucracy was shaped by the historically contingent dynamics of local state structures and practices that were in turn conditioned by but not reducible to the interests and agendas of planned development regimes of twentieth-century India.

The instrumentalist approach to water that characterized the twentieth-century state has produced lasting implications in contemporary India. In keeping with the model of the developmental state, agricultural irrigation has remained the major sector that has drawn on water sources. However, this sector has now begun to compete with increasing demands from urban India. Both the state's instrumentalist approach to water and the complex forms of local centralized control of water continue to shape water governance in the postliberalization period even as the postliberalization period has produced new and distinctive challenges. This has led to an intensification of practices such as groundwater extraction that were key policies of the developmentalist state and the global United Nations Development Programme (UNDP)-oriented models that funded and supported such practices. Local water bureaucratic practices have contributed to the centralization of state authority alongside and in conjunction with mechanisms of central governmental control. The absence of central government regulatory frameworks also provided institutional gaps that facilitated the concentration of local state governmental power through organizations such as the PWD. Such instances illustrate the ways in which processes of decentralization in the postliberalization period provide a set of institutional mechanisms that shift but do not dislodge the concentrated nature of state power in India.

The underlying historically contingent political, economic, and institutional structures of the water bureaucracy are reworked in new ways by successive phases of reform that have been implemented in the context of changing global, national, and local ideational and policy frameworks on the governance of water. The historical continuities that this chapter has foregrounded should not be read as a conflation between the practices and patterns of colonial and postcolonial state rule. Rather, the purpose has been to examine the historical legacies of institutional practices, policies, and cultures that have had lasting implications for the governance of water in India. The legacies of the colonial state are often an understudied dimension of contemporary social scientific studies of reforms in twenty-first-century India. This does not imply that the developmental state has been a lesser factor in shaping contemporary India. On the contrary, the political, economic, and institutional structures of the developmental state are deeply embedded in India. Historical processes provide the backdrop for an understanding of the contemporary liberalizing state and for delineating more precisely what has and has not changed in the context of India's reforms. This context provides both the empirical and analytical space for an evaluation of policies and rhetorical languages of decentralization that purport to break from older models of state centralization and that are now commonplace features of contemporary India.

CHAPTER 2

The Regulatory Water State in Postliberalization India

INDIA'S GOVERNANCE OF WATER IN THE POSTLIBERALIZATION PERIOD has been shaped by two of the central policy tenets that are associated with economic reforms—decentralization and privatization. Both national policy frameworks on water governance that have been emerging since the late 1980s and global projects primarily associated with the World Bank have emphasized these tenets. Local state governments have also begun to promote these norms of decentralization and privatization through the formation of Water Users' Associations and through the restructuring of water bureaucracies. Indeed, critics of reforms in India often focus on the threat of privatization in the water sector.[1] Despite the rhetorical shift that has emphasized privatization and decentralization, shifting policy and institutional frameworks have produced subtle but significant forms of state centralization that shape the governance of water in India. These processes of centralization, which are distinctive features of the postliberalization period, intersect with and reconstitute the historical legacies of state centralization in contemporary India.

Regulatory reforms have been produced by the interplay between global, national, and local water policies and patterns of investment. In the post-1990s period of reforms, the World Bank, the major global player shaping water policies and reforms, shifted its focus from the direct financial support of large water-related infrastructure projects. While antipoverty projects remained a part of the World Bank's work, the Bank shifted to an emphasis on supporting state institutional reforms and private investment for water-related projects often through the structure of public-private projects. The shift to a focus on institutional reforms deepened the interaction between the Bank's policy prescriptions and the frameworks of governance of the Indian state. For instance, there has been a strong convergence between global norms and India's national water policy frameworks and approaches to water governance.

The nature of the convergence between global norms and national policies does not reflect a simple linear relationship where global frameworks were unilaterally imposed on India or evidence of a weakening state in the face of global institutions. India's state has historically had a strong and independent role in its dealings with the World Bank (Prabhu 2017). Rather, state power has been reconstituted and reconsolidated through reforms. Frameworks of institutional reform have consolidated or further expanded the assertion of governmental authority over water resources. In particular, the consolidation of state power is embedded within policy frameworks that are rhetorically associated with decentralization and privatization—processes that are commonly linked with a scaling back of state intervention. In this context, the enduring nature of state centralization cannot be explained solely as a function of the specificities of domestic or political dysfunction in India.

This reworking of state power can be seen in shifts in national patterns of governance and in the specific reconstitution of state power in the case of Tamil Nadu's water reforms. Tamil Nadu represents a significant case for an analysis of the restructuring of the state. The state represents a crucible of national and global trends that have been unfolding in recent decades. Tamil Nadu, one of the most urbanized states in the country, has actively and successfully drawn in private and global investment and has been one of the major (and earliest) recipients of World Bank funding in the water sector. The state has also developed an organization for drawing in finance capital for infrastructure development that has been held up as a national and global model (the Tamil Nadu Urban Finance and Infrastructure Development Corporation). In line with the global norms associated with such funding, the Tamil Nadu state government has engaged in significant institutional restructuring of the water bureaucracy. Given this confluence of patterns, the case provides an important context for understanding the transformation of state power in the postliberalization period.

The architecture of the governance of water that has been shaped by World Bank funding in Tamil Nadu reveals how state authority has been reconstituted through institutional reforms. Institutional reforms have heightened the power of state governmental control and governance of water resources rather than deepening the decentralization of water management. This intensification of local state governmental authority has unfolded in the context of the particular kind of federal structure that has shaped governance over water in India. The federal framework that gave state governments authority over water meant that there was already a significant institutional space for the World Bank to work directly with state governments when it came to water reforms. However, this federalized framework has in turn provided the space for the concentration of new forms of local state authority over water resources.

The World Bank and Global Water Reforms

By the late twentieth century, the World Bank had emerged as the leading global actor shaping norms and policies regarding governance over water resources and water-related infrastructure development. Investment in India reflected this global pattern. The Bank has long been one of India's largest lenders and played a central role in the water sector well before India embarked on its reforms in the 1990s. Infrastructural development in India, as is well known, is one of the major areas that has long been in need of financial investment. The need for the development of basic infrastructure, such as transportation, water, housing, roads, and electricity, has only grown in the context of continuous and accelerating pressures sparked by growth and urbanization. In the face of significant financial needs for this array of sectors, water and sanitation services have represented a segment of infrastructural development that has been one of the least attractive sites for private investment. This risk aversion of the private sector toward investment in the water sector has remained a persistent feature of the postliberalization period in India (TNUIFSL, interview with author, January 2017). The World Bank has stepped in to fill this vacuum both as a consequence of its own priorities and in response to requests from the state in India. The Bank has therefore emerged as the biggest external donor in the water sector in India and has been the leading global actor in shaping India's water reforms (Briscoe and Malik 2006; Prabhu 2017).

In the postliberalization period, the World Bank has shifted from the direct investment in water-related infrastructure (such as its lending for irrigation) in India toward promoting institutional reforms centered on the principles of decentralization and privatization. This shift was part of a general change within the World Bank itself as it moved from an approach that was focused on supporting para-statal institutions from the 1960s to the 1980s to an emphasis on these principles of reform since the 1990s (Bakker 2010, 69). As we will see, these reforms have converged with new policy frameworks of the Indian central government. These policies of decentralization and privatization have been implemented through institutional and financial structures that have reconsolidated the power of the central government. Frameworks of decentralization have paradoxically produced new centralized spaces of state power in the midst of a vast set of smaller and weaker local governmental bodies. Meanwhile, the World Bank has also opted to work directly with local state governments in funding and promoting institutional reforms in the water sector. This work with local state governments has been a natural corollary to its promotion of decentralization. However, as with national policy frameworks, World Bank initiatives have inadvertently intensified new spaces of centralized state authority in a larger decentralized landscape of small and weak local urban and rural governmental bodies. Such reforms, as the chapter will illustrate, provide a vital arena for an understanding of how state power is remade and recentralized at the local and national levels through policies and reforms that are designed to expand decentralization and privatization.

India's water reforms that have been implemented through such shifting local, national, and global norms and policy frameworks provide a distinct case of analysis that illuminate shifting comparative and global trends. In particular, the focus on the centrality of the state in India provides an important counterpoint to the emphasis on privatization that has often characterized research on water reforms. Research in comparative contexts has focused on the deleterious effects of models of privatization (Bakker 2010; Morgan 2011) that have been associated with dominant transnational models of water governance. Indeed, social movements in a wide range of countries including in India have also been focused on their opposition to existing or prospective forms of privatization (Morgan 2011; Urs and Whittell 2009). However, the dynamics of privatization do not adequately capture the nature and implications of this reform model. The centrality of the World Bank's role in shaping water reforms in India itself stems from a deficit of private investment in this sector. As one World Bank report noted, a combination of structural features of the water sector (such as the high capital intensity and a heavily decentralized market) and political risks (such as the political pressures to keep tariffs low) meant that by the end of the first decade of reforms, the water sector accounted for only 5 percent of total private investment in infrastructure (Baietti and Raymond 2005, 1). The Indian context has been a more representative case of the challenges of gaining private investment in the water sector than countries where privatization was rapidly implemented.[2] In India, the complexities of the physical nature of water infrastructure and the institutional landscape have made it a less attractive avenue for private investment in contrast to other sectors of the economy. This has meant that the Indian state has remained a central actor in the management of water resources in ways that are more representative of global trends.

More significantly, the reworking of state authority in India has converged with and is reflective of an underlying shift in the World Bank's own approach to the role of the state in the water sector. In line with an emerging post–Washington Consensus agenda, the World Bank's focus on institutional reforms has identified state accountability as a key component of both its policy ideational frameworks and the specific projects it has funded. These frameworks contain within them the nodes of state authority that intersect with and provide the means for new forms of state control of water resources. In this context, historical forms of state centralization have been reworked but not displaced through reforms that seek to promote decentralization and privatization. The World Bank's shift toward emphasizing state accountability was itself shaped by its experience of the well-known controversies over its funding for the Narmada dam in India. Thus, the Indian case also provided a critical basis for the World Bank's broader rethinking of its approach to the water sector and holds important insights for comparative and global understandings of regulatory reforms of water governance.

The World Bank, the State, and Water Reforms in India

The Bank's work in the water sector in India has a long history that can be traced back to investments in support of irrigation and agricultural development. The World Bank's lending to India first expanded in the 1960s with the establishment of the Bank's International Development Association in

1960. Its investment in water-related infrastructure significantly expanded in the 1970s as the Bank's funding priorities dovetailed with antipoverty programs (Prabhu 2017, 97). In these early decades of lending, the Bank focused on the direct support of infrastructural projects with an emphasis on the irrigation sector. By the end of 1990, the Bank had supported sixty-five projects in the water sector, of which forty-two were for irrigation (WB 1995b). The Bank's support of agricultural development also had more indirect effects on water governance. For instance, the involvement of the World Bank in supporting the model of agricultural development that grew the Green Revolution also meant that the Bank was supporting a form of development that was in large part based on the use of water-intensive high-yielding seeds (Prabhu 2017, 145). The significance of the World Bank as a leading global actor in India's water sector is underlined by the fact that it significantly surpassed funding and investment from other international actors. As journalist Nagesh Prabhu has noted, "Among the foreign borrowings for irrigation for India, the Bank was the major source—71 per cent followed by OECF [Overseas Economic Cooperation Fund]—16 per cent, European Commission (EC)—6 per cent, Germany—3 per cent, Canada—2 per cent, Netherlands and France—1 per cent each in the 1990s" (Prabhu 2017, 153). The importance of the ideational, institutional, and policy frameworks of the World Bank in India are borne out by these patterns of financial investment.

While the World Bank's role in India has continued, the orientation of water-related Bank lending has undergone important shifts in the postliberalization period. The Bank has shifted from an active role in shaping irrigation development in the late twentieth century to a scaled-back role that sought to work through institutional frameworks of the Indian state. With this shift, the World Bank has sought to ensure the state's ownership (and accountability) for the development and change spurred on by the projects that it was funding. The significance of state accountability in this global approach has often been hidden by the heavy discursive emphasis on decentralization and privatization. Such changes have enabled a reworking of the Bank-state relationship in the water sector and have provided the foundation for a return of the Bank to a new active role through its collaboration with the state.

The World Bank's shifts in its approach to investment in India's s water and sanitation sector were in many ways a response to some significant failures in the projects it had supported. Both high-profile political opposition to the World Bank in India and internal Bank reviews critical of its projects led to a significant shift in the World Bank's approach to investment in water-related projects. On the political front, the World Bank's financial support for the Sardar Sarovar Project, designed to build a dam on the Narmada River in Gujarat, provided a central impetus for this review (WB 1995d). As is well known, the now infamous infrastructural project was to provide water for irrigation in drought-prone areas of Gujarat and generate hydropower. The lack of attention to both the human and the environmental costs of the project generated one of the most high-profile antidam movements, the Narmada Bachao Andolan, at the national and global levels (Baviskar 2004; Khagram 2004). The World Bank approved funding for the project over a ten-year period in 1985. However, in the face of sustained protests, the Bank's president initiated the first ever cancellation of a project that was already under implementation (WB 1995d). The report, which concluded that the Bank had not addressed either the human effects of the dam (on farmers or tribal groups who would be displaced by the project) or the environmental effects of the projects, led to the Bank's unprecedented act of withdrawing financial support for the project in 1993 (Morse and Berger 1992). As an independent evaluation group of the World Bank would note, "The Narmada projects have had a far-reaching influence on the Bank's understanding of the difficulties of achieving lasting development, on its approaches to portfolio management, and on its openness to dialogue on policies and projects" (Morse and Berger 1992). In addition to the Bank's realization of the need to explicitly address the centrality of the social and environmental dimensions of projects it was funding, the Bank's Committee on Development Effectiveness concluded that "government 'ownership' should be assured, and social and environmental assessments should be completed, before a loan agreement is signed" (Morse and Berger 1992).

This high-profile review and self-assessment of the Bank was also accompanied by less visible reviews of funding for irrigation-based projects in India that concluded that the impact of such projects had been "less than predicted" (WB 1995a, 1). Consider, for example, the Bank's major "National Water Management Project," funded from 1987 to 1995 with a $176.1 million loan. The project was aimed at improving agricultural productivity and farm incomes through the enhancement of irrigation systems in eleven states. Internal Bank reviews document a range of problems with delays in infrastructure construction and problems with land acquisition and rehabilitation (WB 1987) that led to the Bank rating the project outcome as unsatisfactory. Technical problems produced by construction in some instances harmed segments of farmers, leading them to destroy the infrastructure. As one internal Bank report noted,

> Scheme preparations are not of a high standard. Design and conceptual weaknesses were evident in some of the schemes visited by the mission. In Tamil Nadu, the designer has taken advantage of existing drainage courses in sub-catchments to convey irrigation flows to tanks but in designing control structures for flow division the natural run-off and floods usually handled by such courses have been ignored. As a result, water has backed up on farmers' fields (inadequacy of the structures) and the annoyed farmers have destroyed the flow division structures. No attempt has been made to review the situation and rebuild the structures. (WB 1997, 11)

In other cases, farmers resisted the imposition of agricultural change through democratic processes. For example, a Bank-funded irrigation project that sought to attempt to change cultivation from paddy to dry crops in Andhra Pradesh failed as farmers resisted the transition and obtained a court injunction to stop the project (WB 1997, 6). These internal reviews led to a significant shift in the way in which the World Bank approached its support of investment in India's water sector.

This set of assessments in India contributes to the findings of transnational and comparative research on the World Bank's water projects in important ways. The urban water supply and sanitation projects funded by the Bank from the 1970s to the 1980s received highly negative internal self-assessments and was one of lowest-rated sectors (Bakker 2010, 68). The internal reports on the Indian context reveal that such negative reports were not limited to urban water supply projects but included a broader category of water-related projects. The Bank moved away from direct support for water-related infrastructure projects and instead focused on policy change and improved water management in India. This was evident, for instance, in the Bank-supported reform project, the Tamil Nadu Water Resources Consolidation Project, launched in 1995 (WB 1995b).

The Bank's institutional shifts signify the contradictory nature of the relationship between the Bank and the state that has shaped policies in the water sector in India since the mid-1990s. The Bank's emphasis on policy change since the mid-1990s was concentrated on three significant dimensions—institutional reforms, the financial viability of water utilities and governance structures, and the support for technical improvements that would aid the management of water (WB 1998b). This set of reforms was centered on the Bank's dominant ideational model, which attempted to make space for private sector actors and promote decentralized governance. For instance, the Bank's discursive framework was clearly focused on a model of the regulatory state in which the government was shifting from "provider to facilitator" (WB 2001, 3). However, this model was simultaneously intertwined with the Bank's desire to ensure the government ownership of infrastructural development and reforms based on its own negative experiences in the water sector. The Bank pointed to favorable evidence of this trend in the Indian government's pilot reforms that were designed to change the approach of "providing water through centralized state water boards by sanctioning Rs. 2,500 crores (U.S. $550 million) over three years for piloting water and sanitation reforms in 63 districts in 25 states in India" (1). The World Bank, in effect, placed itself in a supportive role of a government-owned reform process. Noting that the Government of India "has welcomed the assistance of WSP-SA [Water and Sanitation Program–South Asia]," the Bank cast itself as seeking to "support sector reforms by providing proactive implementation and capacity support as well as knowledge management" (4).

In the aftermath of the internal self-assessments of unsatisfactory projects, the Bank also shifted to a new generation of irrigation projects, the Water Resources Consolidation Projects, which would incorporate the principles of institutional reform, technical modernization, cost recovery, and farmer participation (WB 2001, 15). However, once again, underlying these principles was the key foundation of government ownership of the reforms being advocated. The Bank carefully selected its project sites for this generation of reforms in states (such as Tamil Nadu, Haryana, and Orissa) whose governments were actively embracing and supporting its principles of reforms and where the Bank had a prior history of working effectively. The result was a significant decline in World Bank funding in India's water sector (Briscoe and Malik 2006, 71). As one Bank report would note, the Bank avoided projects that would be "reputationally risky"—there was no lending for hydropower, and there were "sharp reductions in lending for irrigation, urban water supply and stand-alone water resources projects, with the only increases being in the uncontentious area of rural water supply. There was great dissatisfaction among government officials in India who believed, as did developing countries throughout the world, that the Bank was walking away from the area where the needs were great (infrastructure) and where the Bank had a strong comparative advantage, namely in addressing complex, difficult issues such as water resources development and management" (Briscoe and Malik 2006 71).

The new World Bank approach continued to emphasize strategies of privatization either explicitly, through the support of privatized or public-private models, or implicitly, through principles of cost-recovery based on an economistic consumer-based model of water supply. However, the Bank's shift to a focus on institutional reform also reflected its view that state institutional frameworks had to serve as the central foundation for the implementation of these normative principles and models. Such a shift again was sparked by the Bank's assessment of failures in the first generation of public-private models in the 1990s. As one World Bank analysis of this early set of partnerships would note, "Most failed because of poor enabling frameworks for private investment, poor project preparation, weak financial strength of project proponents, and opposition to private sector participation" (WB 2014, 1). Such early failures once again consolidated the Bank's shift toward an approach that centered state institutional accountability and change in its efforts to promote its desired water reforms.

While the Bank's focus on water reforms has been focused on both the national and the local levels, India's federal structure has served as the key vehicle for the institutional changes that the Bank supported. This has been in keeping with its overall funding strategies in the postliberalization period as well as in keeping with its policy norms in support of decentralization. As journalist Nagesh Prabhu has noted, "During 1998–2008, out of 107 loans sanctioned by the Bank, 72 loans were granted to [local] states, which constituted 67 percent of the total lending to India" (2017, 345). The World Bank's Water and Sanitation Program–South Asia noted that one of its key successful initiatives was the Cochin declaration, a statement of support for water reforms outlined in India's eighth five-year plan (1992–97) from ministers (WB 2001, 7) participating in the first state water ministers' workshop on rural water supply policy reforms in India, held December 7–8, 1999 (WB 2001). The meeting brought together eleven state ministers, senior central and state government civil servants, NGOs, and representatives of international funding agencies. While the workshop agenda emphasized that the Bank was concerned with social themes of participatory development and decentralization that addressed rural poverty and women's participation, it also highlighted the Indian government's new principle of "water being managed as a commodity and not as a free service" (WB 2000, 2). The framing of the workshop thus rested on a foundation of reforms that were promoted by the central government, with a pledge by local state governmental ministers to implement these reforms. In line with the Bank's emphasis on government ownership of development and reforms in the water sector, the self-representation of the Bank's agenda in this context was one of an enabling rather than an interventionist actor providing "capacity building to strengthen rural water supply institutions and knowledge management" (WB 2000, 2).

Building on this transition in the framework of its support, the World Bank's shift to Water Resources Consolidation Projects, which were designed to work through local state governmental mechanisms, provided a key means to further its reform model in ways that would avoid the reputational risks of large-scale projects such as the Sardar Sarovar Dam project. Such reforms have been evident in a range of Bank-funded state-level water projects. Tamil Nadu's Water Resources Consolidation Project emerged as one of the earliest examples of institutional restructuring of the state's water bureaucracy. Meanwhile, the Bank required the Madhya Pradesh state government to draft legislation for a state water regulatory tariff commission as part of its provision of a $394 million loan for the Madhya Pradesh Water Restructuring Project (Cullett 2009, 90).

Or, to take another example, consider the shift in the model of public-private initiatives in the water sector. The failure of the initial phase of public-private initiatives in the 1990s led to a shift toward more limited projects focused primarily on service delivery. However, the deeper transformation that occurred in subsequent decades was the development of more active governmental authority over such projects. This ranged from the development of governmental regulatory structures designed to support the establishment of such initiatives to the expansion of public funding for such projects.[3] Consider the findings of a comprehensive World Bank study of all public-private partnership (PPP) initiatives in urban water supply with a citywide distribution that were established between 2005 and 2011. Private investment in the five major projects (set up in Maharashtra, Karnataka, and Madhya Pradesh) ranged from 0 percent to 50 percent (WB 2014). Two of the projects had no private investment, one had 10 percent, one 30 percent, and only one had 50 percent private investment. Moreover, as the Bank study noted, private investment shifted from international to domestic capital (WB 2014, 9).

Such trends underline the fact that global dominant norms of privatization and decentralization regarding water sector reforms are being structured by state institutional frameworks and domestic capital and are not simply implemented in a unilateral or straightforward manner either by global financial institutions or by transnational corporations. The underlying objective of the Bank has been to harness state institutional frameworks in ways that will promote its norms of accountability, financial viability, and the growth of the private sector's role. As one Bank report would note,

> One need is to have an institutional apparatus for inter-sectoral water planning, allocation and management. Appropriate institutions for this, comprising a multi-sectoral state Water Resources Board and its State Water Planning Organization, are discussed in the WRM's Report on Inter-sectoral Water Allocation, Planning and Management. Also often desirable is to create a separate state regulatory apparatus, possibly comprised of two entities. One would handle regulation of resource management, in particular of groundwater and surface water abstractions and possibly pollution control. The other would focus on pricing and safeguarding monopolistic practices by water suppliers and users. An immediate need is to establish a water pricing committee which should be independent of political decisions. Over time, as WUAs (Water User Associations) develop and the private sector increasingly enters into water sector investment and management, this body needs to take on full regulatory powers. (WB 1998b, 29)

The Bank's emerging model in the late 1990s thus shifted in significant ways from a simple advocacy of privatization to a more complex model in which the planning and regulatory frameworks of the state would ideally provide both the framework and the foundation for the gradual entry of private sector investment in the water sector.

The question at hand then is how state accountability has been framed through such regulatory reforms and the norms of governance that have been associated with them. The implementation of water reforms in India has produced contradictory sets of processes. The global normative model of the World Bank that has emphasized decentralization and privatization has had a significant impact on water policies and institutions in India (Cullet 2009). This global ideational model has become entrenched within institutions at the national and local levels and has also shaped the ideologies of protest movements and NGOs within civil society (which have focused on the threat of privatization) (Anand 2017).

However, while such changes are important and significant, the underlying emphasis on state accountability has in practice facilitated the creation of new forms of state centralization at both the national and local levels. This veiled centralization is not, as we will see, simply the product of the Indian context corrupting or constraining the ideal-typical global model. Rather, the Bank's model of reform has itself produced key nodal points that facilitated the state's centralization of water resources; this incipient centralization was an intrinsic, if unforeseen, part of the Bank's reform model. Furthermore, such emerging frameworks of reform also took shape within long-standing state institutions and the realities of historically produced political and economic relations and constraints in India. The underlying state-oriented framework therefore provided avenues for the state to rework its centralized control over water resources at both the local and national levels. That is, the newly reformed institutional structures contained within them the spaces for the expansion of state power, even as they were designed to promote decentralization and privatization. The challenge of understanding the implications of global norms of water sector reforms lies in investigating the impact of such reforms on state power at the national and local state governmental levels. Each of these scales of analysis reveals contradictory dynamics, where water governance is liberalized along the ideal-typical model of decentralization and private sector participation while simultaneously centralizing and intensifying state control over water.

Global Norms, National Policies, and the Exercise of State Power through Water Reforms

Institutional reforms in the water sector depart from the conventional narrative of postliberalization economic change, in which the centralized planning of the developmentalist state leads to the increasing devolution of power to the states in the context of India's federal structure. In the early decades of independence, the developmentalist state was marked by weak regulatory frameworks governing water. State governments preserved their authority and become central players in the management of water. Paradoxically, the national planning for water emerged alongside and as a product of India's opening up to global norms of reform. The global model of water reforms would of course remain centered around the dual idealized principles of transforming the role of the state to a facilitator and regulator and increasing the role of both private sector actors and local decentralized organizations. However, the implementation of such ideational and policy frameworks has tended to be implemented through top-down mechanisms. These discursive dimensions of institutionalism foreground the centralized nature of the diffusion of this model of water reforms. In other words, the implementation of these institutional principles as the new normative discursive model rested on a highly centralized process, particularly given the World Bank's new emphasis on clear governmental ownership of all processes of reform.[4]

Consider how such processes of reform unfolded at the national level. First, the central government began developing formal national water policies, in contrast to the early decades of developmental practices that rested on a default acceptance of the constitutional framework that gave state governments primary control over water resources. Second, the postliberalization period has been characterized by major central government developmental initiatives that have combined state investment in water infrastructure with processes of reform. Finally, the central government has also established new national regulatory structures that govern the management of water resources. While such processes of reform often reflect the World Bank's global norms on privatized and decentralized water management, they also paradoxically set into place nodal points for state power that facilitate an expansion of centralized state control over water resources.

Consider first the emerging national water policy frameworks in the postliberalization period. India's first national water policy represented a preliminary attempt at presenting a set of national guidelines for the governance over water resources that accompanied early stages of liberalization that were initiated under Rajiv Gandhi's government (MWR 1987). This early policy framework combined long-standing developmental goals and models of resources management with some of the emerging global norms advocated by institutions such as the World Bank. The policy framework clearly emphasized the need for the provision of drinking water, the need to address the concerns of marginalized groups, and various environmental and public health issues. Such principles were integrated with long-standing developmental models that have stressed modernization and planning. Along with long-established developmental principles of equity and access, the national policy framework also articulated newer languages that reflected global norms of a model that incorporated a framework of participatory and decentralized management on the one hand with the establishment and collection of water rates on the other. As the policy noted, "Water rates should be such as to convey the scarcity value of the resource to the users and to foster the motivation for economy in water-use. They should be adequate to cover the annual maintenance and operation charges and a part of the fixed costs. Efforts should be made to reach this ideal over a period, while ensuring the assured and timely supplies of irrigation water. . . . Efforts should be made to involve farmers progressively in various aspects of management of irrigation systems, particularly in water distribution and collection of water rates" (MWR 1987).

In later phases of the postliberalization period, this integration of new global norms within the policy-planning framework was significantly expanded in a newly reworked national water policy. In a striking parallel to the World Bank's shift away from direct infrastructural investment to a focus on institutional norms, the 2002 policy contained new tenets advocating institutional reforms. The policy stated that "the existing institutions at various levels under the water resources sector will have to be appropriately reoriented/reorganised and even created, wherever necessary" (MWR 2002). In addition, the policy presented the first official national statement explicitly encouraging both private sector participation and processes of decentralized participatory management through Water Users' Associations (in ways that once again converged with the World Bank's global norms).

It is commonplace to view this pattern of devolution of state power through processes of decentralization as a shift away from centralized state intervention toward more local state autonomy. Indeed, in the postliberalization period, this devolution has been formalized through governmental legislation that has sought to enhance local governance in both rural and urban communities (Panchayati Raj Institutions and urban local bodies) by expanding their governance and financial authority.[5] The implications for the water sector were significant, as they were intended to place the responsibility for service provision on ULBs.[6]

Such shifts in the postliberalization period along with the weight of public and political rhetoric on local governance have often elided an understanding of how decentralization in fact produces an intensification of centralized forms of state power. While there are variations in the models of decentralization in arenas such as governance of rural water supply (S. Singh 2016), there has been a dominant model of reforms that has been promoted through centralized mechanisms. National policy frameworks and decisions have reflected a strong degree of nationalization and centralization in the water sector (Warghade and Wagle 2011, 328).

Consider, for instance, the ways in which water reforms were promoted by the central government. Central government financing of local projects has increasingly been tied to conditionalities that state governments demonstrate that they are meeting central governmental norms of reforms. An early example of this was evident in the formulation of new national guidelines for the provision of rural drinking water in the form of the 2002 Swajaldhara Guidelines, which were formulated at the same time as India's new 2002 national water policy framework (MRD 2002). The new guidelines specifically sought to shift "the role of Government from direct service delivery to that of planning, policy formulation, monitoring and evaluation, and partial financial support" (MRD 2002). The new model of water provision reflected the new global-national consensus on local governance and cost recovery. However, the implementation of this model rested on the financial control of the central government. In the initial phase, the new model delineated "up to 20 percent of the Budget provision for Rural Water Supply Programme of Government of India" (MRD 2002, 9) from the tenth five-year plan for projects that met the new reform guidelines. This conditional linking of central government financing with state-level reforms has continued to expand. For instance, the main policy change in the Accelerated Irrigation Benefits Programme in the twelfth five-year plan included "Enhancement of Central Assistance up to 50% for ongoing and new projects of General Areas subject to the States carrying out water sector reforms and satisfying the 'Reform Friendliness' benchmarks" (MWR 2014, 4).

The financial structure underpinning this process of decentralization shifts the accountability of governance (for the adequate and effective provision of water to communities) to local *panchayats* and urban local bodies (ULBs). However, *panchayats* and ULBs are not given the financial autonomy and remain dependent on both state and central governments. For example, between 2007–8 and 2012–13, "there was an erosion in municipal financial autonomy across the country. The smaller the size of the ULB, the greater its dependence on intergovernmental transfers to finance civic services and facilities" (P. Mohanty 2016, 22). Similar trends are prevalent in the case of *panchayats* (P. Mohanty 2016, 24). The fourteenth Finance Commission report confirmed that "the representatives of panchayats and municipalities in an overwhelming majority of States mentioned that they faced a paucity of funds for carrying out their own mandated functions" (Reddy 2015, 102). As political scientist S. N. Sangita has further noted, local governments "still depend upon higher level governments for about 70–80 percent of their expenditure" (2014, 91). Decentralization has in effect encoded local governments within a framework of centralized control through such structures of financial dependence.

Meanwhile, the new dominant centralized discursive model of water governance was reinforced in the commission's recommendation that "states (and urban and rural bodies) should progressively move towards 100 per cent metering of individual drinking water connections to households, commercial establishments as well as institutions" (Reddy 2015, 213). The enactment of reforms produces an increased financial precarity of local rural and urban bodies that enables the enforcement of top-down decisions that are characteristic of a centralized model of water governance. While the delegation of authority to local *panchayats* and urban local bodies is generally cast as an endeavor designed to enhance local governance and authority, the underlying financial dependence on the central and state governments in fact intensifies the centralization of state authority.

This pattern of the centralization of water governance converges with parallel patterns of centralized control that stem from the developmentalist state. What is distinctive about the postliberalization state then is not a shift from centralized state power to a process of decentralization that enhances local state authority but a shift in the arena of the state's accountability for the provision of services. The central government has in practice either maintained or expanded its control over water resources in the postliberalization period, while the responsibility for providing adequate water supplies to citizens and other sets of consumers has shifted to local governments. What is decentralized, then, is state accountability to citizens but not the centralized financial control that underpins state authority.

Such dynamics have meant that national regulatory frameworks that are necessary for effective water governance have become enmeshed in these dynamics of centralization. For example, the central government has been encouraging the establishment of a Water Regulatory Authority at the local state governmental level. This framework, which is in keeping with the World Bank's long-standing advocacy of this specific form of institutional structure, is designed to provide a vehicle for the enforcement of pricing; water regulation in this context is not geared toward broader questions of access, equity, or environmental issues.[7] As the fourteenth Finance Commission report recommends, "We reiterate the recommendations of the FC-XIII and urge States which have not set up WRAs to consider setting up a statutory WRA so that pricing of water for domestic, irrigation and other uses can be determined independently and in a judicious manner" (Reddy 2015, 214). Similarly, the state has also established a centralized agency to oversee and facilitate public-private partnerships.[8]

This kind of centralization has been interwoven with regulatory efforts that are necessary for effective governance. In recent years, the central government has attempted to revise its approach to water governance and to produce new regulatory mechanisms. The twelfth five-year plan set in place a paradigm shift that sought to encode new principles of reform, decentralization, and sustainability (Shah 2013). Furthermore, the central government has realized the need for the effective management of groundwater. The Central Ground Water Authority (CGWA) has maintained the authority to regulate and control the management and development of groundwater since 1986. The 2002 national water policy signaled the need for the government's continued public control over groundwater by both the central and local state governments in order to prevent the overexploitation of the resource. More recently, the government developed the Draft National Water Framework Bill, 2016, and the Draft Model Bill for the Conservation, Protection, Regulation and Management of Groundwater, 2016, and recognized the need for the reorganization of the CWC and CGWB (Shah 2016). As with the state-led model of decentralization, the central government has been pressing local state governments to enact groundwater legislation based on its model bills. Such regulatory mechanisms are necessary given the growing stresses on groundwater, but they remain dependent on the role of state governments to both enact and implement the regulations.

However, these new attempts to develop regulatory mechanisms have been accompanied by more conventional forms of state centralized authority that echo the approaches of the developmental state. The command-oriented approach of the twentieth-century state is evident in the major national river-interlinking project that has been undertaken by the Modi-led government, in response to a Supreme Court directive.[9] The interventionist nature and scope of the project signals the continued salience of large-scale developmental projects that rest on conventional modernist visions of infrastructural endeavors that were central features of the early decades of India's developmental state.[10]

This long-standing dynamic of centralization through developmental agendas has taken the form of new state-sponsored programs for infrastructure development. The language of infrastructural development has in effect replaced earlier modernist languages on development. If dams and factories were the infrastructural symbols of the Nehruvian vision of development in India, metropolitan city and large urban centers have become the new emblems of postliberalization developmental discourses. In this context, the reinforcement of state power in shaping the trajectory of the water sector has been shaped by the financial underpinnings of major state-led initiatives designed to promote urban infrastructure. The first major state-led initiative, Jawaharlal Nehru National Urban Renewal Mission (JNNURM), implemented in 2005–14, included water and sewerage infrastructure as a central dimension of its focus.[11] As Piyush Tiwari and Ranesh Nair have noted, by the end of 2010, nearly 60 percent of the spending by JNNURM was in the water and sewerage sector (2011, 240). As with the earlier governmental assistance models, funding provision required mandatory reforms at the level of both the ULB and the state government (12). Furthermore, in accordance with the national water policy framework, the JNNURM did attempt to balance both the mandatory reform of achieving full operation and maintenance costs through user charges with the state's obligation to provide services to the urban poor.[12] The program favored larger and well-off states in India, and the implementation produced subtle forms of centralization by disconnecting ULBs from city development planning; from the preparation, approval, implementation, and supervision of projects; and from the assessment of the public benefits of projects. The program thus expanded the power of local state governments over ULBs (Bhide 2017). Centralization in this process was evident across spatial scales in both the top-down mechanisms of the JNNURM and the concentration of local state governmental authority.

While the BJP-led government discontinued the JNNURM after it came to power in 2014, it replaced the initiative with a new five-year plan, under the framework of the Smart Cities Mission for the period 2015–20. The Smart Cities Mission also incorporates water and sanitation infrastructure as a primary component for funded projects. Furthermore, the new Mission reflects an extension of the existing framework of reforms by requiring matching funds from state governments and ULBs and the mobilization of private investment to supplement state funding. Given the financial strains on most ULBs for the provision of basic services, such funding conditionalities exacerbate inequalities between metropolitan cities and wealthier towns on the one hand and smaller, poorer urban localities on the other. The competitive framework of the Mission that requires cities and ULBs to compete for the funds further accelerates the race to the top for wealthier localities and allows for potential political considerations that shape center-state relationships to structure access to infrastructural funds. These dynamics are of course familiar ones that are associated with the planned economy of the twentieth-century developmental state.

Such initiatives are significant as they both highlight the continued centralized state governance over water and caution against easy assumptions of a clear transition between the model of a state-managed regime over water resources and new models of decentralization and privatization in the post-liberalization period. As a World Bank report noted,

> Since 2005, a growing number of urban water supply PPP projects have been developed on the basis of substantial public funding. At present, 50 percent of projects have been developed with financial support from the central government. The capital injection from schemes such as JNNURM and UIDSSMT [Urban Infrastructure Development Scheme for Small and Medium Towns] has been a major driver of this shift. Public funding for PPP projects in progress within the JNNURM framework (including the UIDSSMT component) covers approximately 60–70 percent of the escalated project cost. . . . Given the high risk perceptions about water PPPs in India, the share of private investment is likely to remain limited, and reliance on public funding substantial. Moreover, given the weak financial health of ULBs, most public funding would need to come from state and central government sources, rather than ULBs. (Swaroop 2011, 8)

Or consider another example of how centralization is reworked through the case of Water Users' Associations and models of farmers' participatory irrigation management. In accordance with national-global norms regarding decentralization, most states in India have established Water Users' Associations to deepen processes of decentralization. However, such processes have often inadvertently reproduced older modes of state bureaucratic authority in a range of states' water-related management activities, such as watershed management and large irrigation system management (Baviskar 2004, 31; Manor 2004, 203). In such cases, centralized forms of authority permeate decentralized projects through the local bureaucracies of the state government. Decentralized models of participatory management that have been developed for rural water management are located within a broader institutional structure that is itself being reworked in ways that are weighted toward the state control over water resources through powerful water bureaucracies that serve larger cities.

The significance of state governmental authority is also underlined by the fact that India's constitutional framework placed the management of water resources primarily within the purview of state governmental power. The initial phases of reforms in the water sector were largely driven by specific, localized changes in particular states. States such as Andhra Pradesh, Madhya Pradesh, Maharashtra, Uttar Pradesh, and Arunachal Pradesh made early attempts at reforms. The new national reform-oriented Swajadhara guidelines were based on the local model of reforms implemented in Uttar Pradesh. Meanwhile, reforms promoted by the World Bank Water Resources Consolidation Projects were concentrated in Haryana, Orissa, and Tamil Nadu. State-level reforms provide the central arena for the implementation of the global norms and national reforms in India's water sector, and local processes within states remain the crucial means for understanding the dynamics of institutional reform.

Water and the Centralization of State Governmental Power in Tamil Nadu

Tamil Nadu has had a long history of adopting global developmental models and working with international agencies in developing and managing its resources in the water sector. In more recent years, the state has been at the forefront of implementing reforms in the water sector, and the state has had an extensive working relationship with the World Bank. Given the state's embrace of the new global model of reforms, including both incorporating private sector participation and engaging in institutional restructuring, Tamil Nadu represents a strong case for assessing how patterns of state centralization are reproduced within its institutional models of water governance.

In the early decades of independence, UNDP assistance shaped the state's management of groundwater. Since then, the World Bank has an established history of funding projects in the state. Early projects taken up from the 1980s ranged from the development of drinking water supply infrastructure in both rural areas in the state and in major cities such as Coimbatore and Madras (Chennai) to as a major restructuring project in the state's irrigation sector, the Water Resources Consolidation Project (WB 1995a, 1995b, 1995c). The Water Resources Consolidation Project in particular represents an exemplary case of the Bank's shift away from infrastructural funding toward institutional restructuring, and the project represents one of the central examples of the Bank's new approach to the water sector.

Tamil Nadu's Water Resources Consolidation Project, for instance, was an early example of the implementation of the principles of institutional reforms, cost recovery, and farmer participation (WB 1995a). As the World Bank memorandum recommending a $282.9 million loan would note, "In preparation for the project, and following on from its 1994 State Water Policy, GoTN has commenced a rigorous program of policy and institutional reforms" (WB 1995b, 3–4). These reforms included a wide range of changes, such as the creation of a specialist Water Resources Organisation (WRO), staff reorganization, programs designed to increase farmer participation, and procedures for the annual review of cost recovery and water charges. As the Government of Tamil Nadu's letter requesting the loan would note, Tamil Nadu had already made significant advances in raising charges for water usage by farmers (November 1, 1994). The state's charges have been, as the letter would note, one of the highest in the country, where "rates for bulk water supply for industrial and commercial use were increased by over six-fold in 1991, and agricultural rates have been periodically adjusted through increases in the irrigation cesses. An opportunity is also present, through farmer organizations, to progressively internalise revenues collection and expenditures at the levels of the operating systems and to eventually adjust to volumetric supply and charging arrangements linked to service costs" (4–5).

In keeping with the Bank's shift to safer investments, Tamil Nadu's demonstrated commitment to its model of water sector management provided a reliable foundation for its self-conceived role as a facilitator of government-sponsored reforms. The Bank's major sectoral review (WB 1998b) of its investment history and practices in India would note that dimensions of such institutional restructuring as well as the reorganization of Chennai's major water utility were positive signs of change. The state represented one of six states (along with Andhra Pradesh, Maharashtra, Madhya Pradesh, Uttar Pradesh, and Rajasthan) that the Bank would select for Water Restructuring Projects after it had conducted its in-depth sectoral review (Burton and Dhingra 2014, 21).

In addition to its systemic support of institutional restructuring in the state, the World Bank has also funded one of the key mechanisms for private investment targeted primarily at urban local bodies. Given that the nature of the water sector has made it a less attractive arena for private sector investment, the ability of international financial institutions to provide risk mitigation for private investment in infrastructure has been critical (Baietti and Raymond 2005). Tamil Nadu's Urban Development Fund (TNUDF) has represented a model both in India and at a global level in its endeavor to provide such mechanisms designed to draw in private capital for infrastructure, including within the water sector. The program was the product of sustained financial support from the World Bank in urban municipal development in the state. The investment, which began in 1988, led to the establishment of the TNUDF in 1996 in partnership with both the Tamil Nadu government and support from central governmental institutions.[13] The fund was the first such public-private endeavor for municipal funding in India and is often held up as a national model for the mobilization of private capital for infrastructural development in the country. As the World Bank proposal for a third phase of funding of $300 million for the TNUDP noted,

> The Second Tamil Nadu Urban Development Project (TNUDP-II) made a very strong impact on urban reform and strengthening of ULB capacity. The Tamil Nadu Urban Development Fund as established under TNUDP-II has been successful in bringing ULBs to the market and exposing them to commercial borrowing practices. Both GoTN [Government of Tamil Nadu] and GOI [Government of India] see the continuation of this collaboration as a way to consolidate urban reforms in Tamil Nadu and to provide sustainability

and continuity to the access to financial markets for urban local bodies. In addition, GOI sees this as a way to bring to fruition a model that could be replicated at the national level and in other states as they reach Tamil Nadu's level of urbanization and implement reforms similar to those that have been implemented there, while GoTN sees it as a continuation of a long and fruitful relationship with the Bank (WBIEU 2005, 4).

Indeed, the fund manager (TNUIFSL) for the TNUDP has developed a strong record in the loans that it has managed. According to its 2019–20 report, the fund reported a 100 percent debt recovery record over a period of sixteen years (TNUDF 2020). The track record of globally funded reforms has meant that Tamil Nadu represents one of the central cases where reforms and global norms of water sector management have been implemented in the postliberalization period.

In practice, the reforms in the state have exemplified the processes of centralization that I have analyzed. For example, processes of incipient centralization have been particularly significant in the case of one of the models of decentralization—the case of Water Users' Associations and committees. Consider the implementation of processes of decentralization in Tamil Nadu. The seventy-third and seventy-fourth constitutional amendments were passed in the state without any discussion in the state assembly (Kumar 2011, 27). This kind of top-down approach has been duplicated in the example of Water Users' Associations that were established by a state governmental order without the preparatory groundwork that could have deepened the effectiveness of this decentralized framework.[14] As one report noted, despite problems with the initial phase of Water Users' Association operations, the government passed a new act, the Tamil Nadu Farmers Management of Irrigations Systems Act, in 2001 "for a new set of bodies linked to WUAs before studying the soundness" of the organizations (CWR 2003, 75). The act reworked the original 1994 three-tier framework into a three-tier Farmers Organisation, consisting of "Water Users' Association (WUA) at Primary level, Distributory Committee (DC) at Secondary level and Project Committee (PC) at the Project level" (WRD 2018). While the state has gone on to hold formal elections for posts on these committees, the system of participatory management is placed within the state's centralized water bureaucracy. Thus, the engineers of the public works department were appointed "as the Competent Authorities for the Water Users' Associations, Distributory Committees and Project

Committees respectively" (WRD 2018). Water Users' Associations were thus incorporated into the lowest rung of the state's water bureaucracy.

Meanwhile, the state also captures the dynamic of the divergence between state centralization and effective national regulatory authority. Such processes of centralization have also been met by resistance or apathy that has stalled institutional and policy changes. Such resistance has taken the form of a competition for state control between the central and state governments. The case of Tamil Nadu illustrates both the effects and limits of this interplay between central and local state governmental power. As we have seen, there has been a nationalization of water policy frameworks that the central government has sought to use as a means of producing a more uniform policy across state governments and more effective regulatory mechanisms. The central government has attempted to encourage state governments to adopt state water policies based on the national water policy framework. Tamil Nadu has been one of the more proactive states attempting to develop water policies that are generally in line with national frameworks. However, while such frameworks have been developed, they are often stalled in the context of local governmental and political processes. The Tamil Nadu government, for instance, developed a water policy that was closely modeled on the national framework in 2007 but was never formally approved by the government (PWD, n.d.b., 2).

Consider a second example of how national regulatory frameworks have become ineffective in Tamil Nadu—the case of groundwater legislation. The central government has periodically circulated model groundwater bills that it has encouraged state governments to adopt. While the circulation of such bills did not originate with the reform period (model bills have been circulated in 1970, 1992, 1997, 2005, and 2016), the growing governmental concerns over groundwater as usage and competition over the resources continues to intensify have accentuated the significance of such proposed legislation. Tamil Nadu has made attempts to adopt groundwater legislation. The Chennai Metropolitan Area Ground Water (Regulation) Act, 1987, was an early example of such regulation and remains in force for the metropolitan area.[15] Meanwhile, legislation covering the remainder of the state was developed through the Tamil Nadu Groundwater (Development and Management) Act, 2003, and enacted in 2004. The context of the legislation was a lack of action at the local state level, where, as one Ministry of Water Resources report noted, "Barring a few exceptional cases, political and administrative leaderships in most states have been reluctant to impose any restrictions for management of ground water. The Model Bill has been in circulation ever since 1970 i.e., for the last 37 years. But there have been very few takers" (Prasad 2008, 74). In the case of Tamil Nadu, the government never notified the bill and then revoked it in 2013.

These new forms of centralization have been interwoven with the renewed interventionalist models associated with the developmentalist state. For instance, the Tamil Nadu government has played an active role in pressing for the nationalization of river waters through the central government's river-interlinking project. When the Supreme Court asked states to respond to the proposed project, Tamil Nadu was the only state to provide a response.[16] When the Congress-led government avoided action on the Supreme Court's order to establish a Special Committee to implement the project, the Tamil Nadu government proactively "requested that the Special Committee should be activated and all Inter-State rivers should be nationalised so that water resources of the country are optimally utilized" (Palaniswami 2017, 101). Since the constitution of the committee under the Modi-led government, the Tamil Nadu state government has continued to press for the immediate implementation of the national river-linking scheme (103).

Such activity by the Tamil Nadu state government has been shaped by Tamil Nadu's problems with acute water scarcity. This problem with scarcity is exacerbated by its geophysical conditions and the constraints from its location downstream from rivers in its neighboring states. As we will see in the next chapter, the state government has also sought central government intervention through a separate set of regulatory institutions that have governed interstate water disputes. This has produced protracted claims on the state machinery of interstate river-dispute tribunals and the Supreme Court. Questions of state reform and regulation in Tamil Nadu thus bring to the fore the two faces of the postliberalization state in India. First, the state has embraced an emerging reform-oriented regulatory state that has been shaped by new global-national dominant norms and produced new forms of centralized power. Second, such patterns of centralization have intersected with long-standing state structures and institutions with autonomous histories and practices.

It has become commonplace for scholars of reforms to emphasize the significance of state governments in initiating and implementing such policy changes. State governments have indeed played an active role in activities including pursuing private investment, obtaining loans from international financial institutions, and restructuring local governments. The significance of local state governmental activity is compounded in the case of water policies, as water has been constitutionally designated as under the primary preserve of state governmental authority. Furthermore, the postliberalization period has been marked by subtle attempts of the central government to increase its control over water resources. Such attempts once again illuminate the underlying processes of centralization that are embedded within new frameworks of decentralization. In contrast to the conventional story of the postliberalization Indian state as an emerging federalized, regulatory state, the case of water reveals a more complex set of centralizing processes at the local and national levels. In the next chapter, I use an examination of interstate water disputes and negotiations to illustrate the complexities of governing water through federalized center-state relations that are structured by the political economy of liberalization.

CHAPTER 3

The Political Economy of Federalism and the Politics of Interstate Water Negotiations

INTERSTATE NEGOTIATIONS OVER THE SHARING OF WATER RESOURCES and water-related infrastructure provide a unique avenue for an understanding of the nature of federalized state authority in postliberalization India. Tamil Nadu's relationships and conflicts with its neighboring states show how the legacies of the political economy of both the colonial and developmental state are reconfigured in the postliberalization period. More recent developmental pressures of economic growth intensify deep-seated historical strains on water governance. The stratified nature of water governance brings to the fore the consequences of an unwieldy blend of assertive local state governments on the one hand and weak national regulatory mechanisms on the other hand. These contradictions in the federalized governance of water produce institutional dysfunctions and gaps that then in turn draw in the participation of centralized institutions, such as the machinery of tribunal structures and the Supreme Court. Furthermore, the institutional terrain of interstate water governance has intensified the political claims of state governments in ways that consolidate local governmental attempts to assert their authority over water resources.

Bureaucrats navigate these complex historically constituted institutional, political, and socioeconomic fields. At one level, as might be expected, bureaucrats consolidate the power of local state governmental authority and harden the lines between competing state governments. However, in other cases, bureaucrats provide crucial space for the resolution of interstate negotiations in ways that unsettle homogenized conceptions of a static and ineffective bureaucracy. Understanding how effective bureaucratic agency can be foreclosed by broader political, economic, and institutional contexts is as critical to understanding deficiencies in governance as identifying moments of successful bureaucratic action.

Consider the following vignette from the professional diary of a major figure in Tamil Nadu's water bureaucracy:

> Ms. Jayalalitha, Chief Minister, just goes to Anna Samadhi, Marina, in the morning and sits there saying she is on indefinite fast to urge the Central Government to take action to implement the Interim Order of the Tribunal.
>
> Though it was conceived as a strategic move, the sudden decision took even her own Ministers by surprise. They all came running to the Anna Samadhi. . . .
>
> Chief Minister holds the Cabinet Meeting with the Ministers standing around her bed in Marina. A strange scene. Hon'ble Shukla contacts Karnataka over phone and he and the Chief Secretary were trying some modification to the draft to manage the situation. . . . Mixed reaction in the Press on the fasting drama. The whole thing has turned out to be a Political gimmick, the results of which are not known. The Chief Secretary calls for a meeting we will attend and he said he is not able to have any agreed text which is still in the drafting stage for the formation of the Monitoring and Implementation Committee for the Tribunal's Interim Order. Entire day spent in Secretariat attending meetings and drafting notes. (Mohanakrishnan 2016b, 59, 61)

The scene depicting the political dramatic acts of then Tamil Nadu chief minister Jayalalitha provides a rare glimpse of the view of interstate disputes from the perspective of a state bureaucrat. While water-related disputes have often been a fraught area of contestation between state governments, much of this conflict is publicized either through intense political conflicts in the public sphere or through the records of legal adjudication (whether through the centralized state machinery of tribunals set up by the central government or by legal suits brought by state governments before the Supreme Court).

Formal records of tribunal and court judgments provide skeletal accounts of legal claims, technical evidence, and judiciary responses between state governments with hardened positions and conflicting interests.

These observations of Chief Minister Jayalalitha's protest unveil a more complicated set of dynamics that surround what we have come to know about the role of the state in such conflicts. The observations, recorded in the professional diary kept by a distinguished state administrator in Tamil Nadu, provide a rare glimpse of the inner workings of the state. They refer to a small slice of events in the dispute between Tamil Nadu and Karnataka over the Cauvery River, India's longest and most volatile water-sharing dispute, which has unfolded through decades of ministerial meetings, adjudication through the official Cauvery Waters Tribunal, and multiple appeals to the Supreme Court.

This inner glimpse encapsulates various intersecting layers of the state. The chief minister's public fast reflects the deep politicization of interstate water conflicts. State officials use such visible public acts to mount pressure on the central government, to win over public sentiment within the state, and to neutralize opposition from competing political parties. Such forms of public theater, which are characteristic of Indian democratic politics, highlight the complexities involved in negotiations over water sharing and the political strains on the federal institutional mechanisms that are designated for the management of water resources. These political contestations are in turn aggravated as the pressures of accelerated and unplanned economic growth are intensified by the increased competing demands on water in times of drought and scarcity.

A second feature of the state that is evident in this vignette is the sluggishness of state action over difficult political issues. The dramatic speed of the chief minister's rush to the Marina for a hunger fast is matched in inverse proportion with the hidden scenes of phone conversations, negotiations over an "agreed text," and the laborious work encapsulated by the simple description "Entire day spent in Secretariat attending meetings and drafting notes." It would be relatively easy to reduce an analysis of the Indian state to a well-known narrative of the dysfunctions of Indian democratic politics, where politicians use political theater to shore up their own political strength, or to the familiar story of a sluggish state bureaucracy that does not have the capacity or will to act on critical political and economic issues.

Yet the observer describing these events, Professor A. Mohanakrishnan, is himself a member of the state administration and at the time of the events was chairman of the Cauvery Cell of the Tamil Nadu state government. Professor A. Mohanakrishnan's personal records of this vignette of the Cauvery dispute provide a window into a facet of the state that does not correspond to familiar one-dimensional stories of politicization, lethargy, or corruption. This dimension of the state has to do with individuals and institutional actors that negotiate an array of political quagmires and institutional hurdles in order to implement policies, reach agreements, and manage resources. Bureaucrats in effect do the public work of the state through everyday practices that are, paradoxically, largely not visible to the public. A focus on such, usually hidden, dimensions of state practices opens up important analytical space that unsettles exceptionalist arguments that identify a monolithic form of bureaucratic stasis or corruption as the overriding source of problems of governance in contemporary India.

Interstate Disputes in Southern India

Tamil Nadu's interstate water-related disputes and negotiations with its three neighboring states of Andhra Pradesh, Karnataka, and Kerala bring together these intersecting facets of the state in a unique way. The political, institutional, and agential dimensions of state action are both shaped and constrained by historical structures of political economy. Colonial and postcolonial patterns of development have produced local and regional political-economic conditions that have in turn sparked the prolonged interstate conflicts over water resources that currently weigh on the states that share water resources and infrastructure. Historical structures of political economy have been specifically reshaped by state-led policies of liberalization as patterns of urbanization and new patterns of investment have intensified competing demands over water resources. Interstate disputes and negotiations involve a range of state and civil society actors including the central and state governments, the centralized institutional machinery of tribunals, the Supreme Court, political parties, and social movements. This breadth of actors involved provides a unique understanding of the complex entanglements of the federalized governance over water in the postliberalization period.

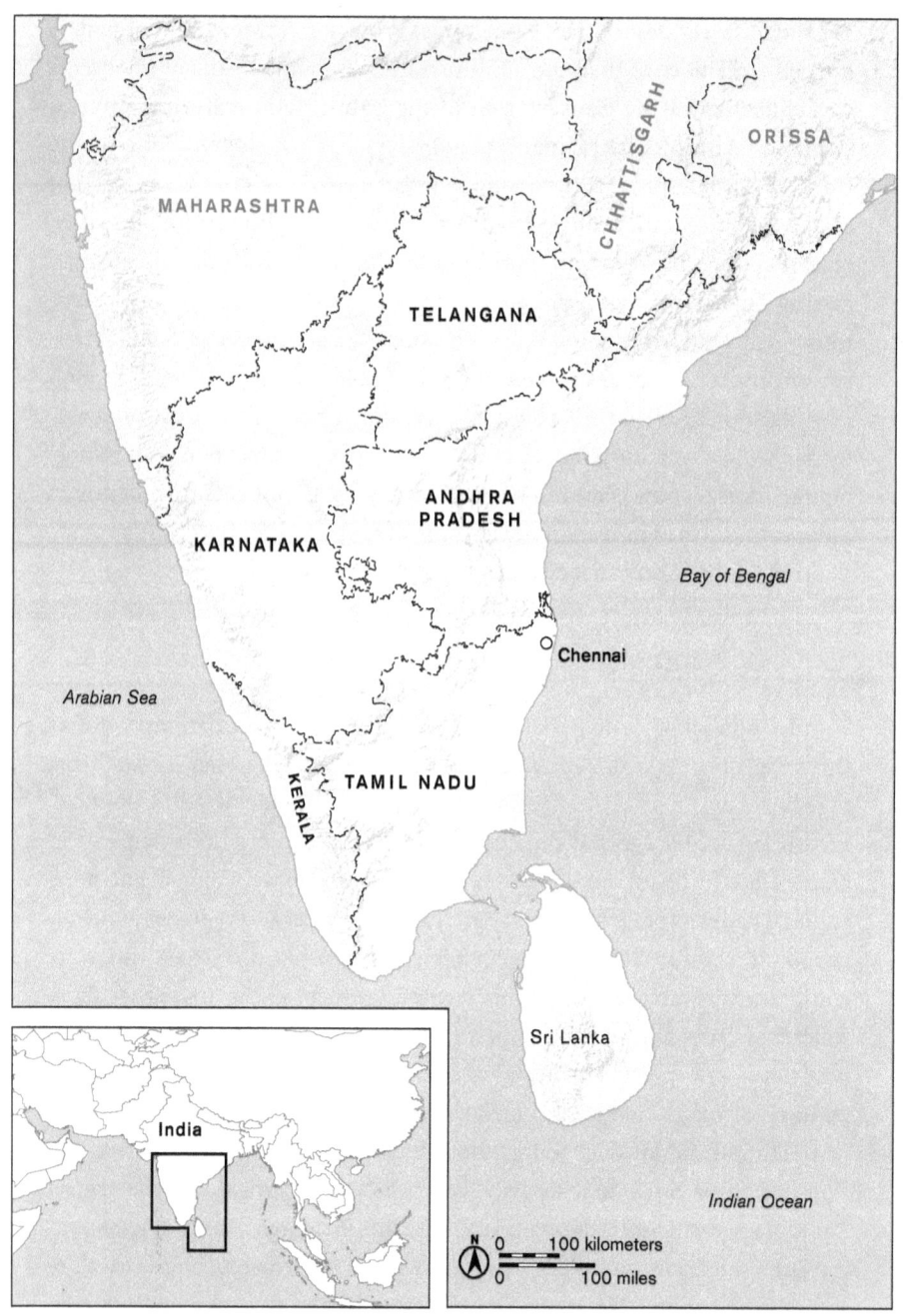

MAP 3.1. Southern Indian States, showing the states that make up India's southern region

Both India's planned developmental state in the twentieth-century and the contemporary postliberalization state have inadvertently produced a federal framework that has exacerbated competition between state governments.[1] When such competition takes place over pliable resources, such as budget allocations or private capital, it can be managed politically or remain relatively invisible (especially when both analytical and political-administrative frameworks take local state governments as an autonomous discrete unit). Furthermore, "India has fourteen major rivers, which are all inter-state rivers and 44 medium rivers, of which nine are inter-state rivers having catchments of watersheds in two or more states" (Padhiari and Ballabh 2008, 174). In most contexts, the routes of rivers run without incident. However, when decades of planned agricultural development and intensified urbanization in the context of a liberalizing economy intersect with nonhuman constraints of drought (and the growing unpredictability of weather patterns in the context of climate change), the salience of addressing relationships between states is underscored.

Tamil Nadu represents a significant case for an understanding of this political economy of federalized water governance in the postliberalization period. Tamil Nadu, a lower riparian state, is reliant on water-sharing arrangements and the shared management of water-related infrastructure with all its three neighboring states. The state has also had a history of drought and periods of water distress that have intensified in the postcolonial period and that have brought the state to periods of severe crisis in recent years. The failed northeastern monsoon in 2016, for instance, created acute shortages of water for both agricultural and urban areas. The result was a series of failed crops, farmer suicides, and dried-up reservoirs that supply water to the city of Chennai. This heightened both governmental and political attention on water that was due to the state from Andhra Pradesh and Karnataka according to two interstate agreements. However, given that Karnataka and Andhra Pradesh were also facing water resource constraints, they each were, in different ways, stalling on the release of waters. While Tamil Nadu was unsuccessful in getting Karnataka to abide by the final judgment of the Cauvery River tribunal award, an emergency trip by Tamil Nadu's chief minister to Andhra Pradesh was at least partially successful in gaining a promise of the release of some water from the Krishna River. The contrasting dynamics of these two examples of interstate interaction illustrate that negotiations between states over water resources and water infrastructure do not inevitably produce intractable conflicts. The dynamics of interstate relations are shaped by historically contingent constraints and patterns of political, economic, and institutional practices that undergird the postliberalization state in India.

The variations between these three cases provide a unique understanding of an understudied dimension of the dynamics of federalism and the complex nature of decentralized state authority in the postliberalization period. Variations between the agreements also point to the importance of understanding how the agential contingencies of political and bureaucratic actors play a significant role in shaping the relative successes or deficiencies in interstate cooperation. The three interstate water-related agreements that link Tamil Nadu with Andhra Pradesh, Karnataka, and Kerala vary in significant ways in terms of the scope of the agreements and the kinds of issues under contention, the political dynamics of the agreement, and the outcome and implementation of each of the agreements. The first case, the Cauvery waters dispute, represents one of India's longest and most politicized conflicts over water sharing between Tamil Nadu and Karnataka. The Cauvery case, which has seen prolonged adjudication in both the Supreme Court and the Cauvery Water Tribunal and intense politicization, including the outbreak of periods of ethnic violence, was officially concluded with a final judgment from the tribunal in 2007, after thirty-seven years of review, negotiations, and conflict. However, the implementation of the agreement remains a continued site of political conflict, particularly in distress years, when both states face severe water shortages. The second case, the Krishna Water Supply Project (also known as the Telugu Ganga Project) represents a negotiated bilateral agreement (spurred by central government intervention) that channels waters from the Krishna River to supply drinking water to Chennai. The agreement is largely seen as a successful case of interstate cooperation. The third case involves a prolonged dispute over Kerala's concerns over the safety of the Mullaperiyar Dam, which is located in Kerala but fully operated by Tamil Nadu. As this case represents a conflict over water infrastructure rather than riparian rights, it played out through a long judicial process in the Supreme Court that was ultimately decided in Tamil Nadu's favor. However, while Tamil Nadu's operational control of the dam removes any practical obstacles to implementation of the judgment, the politicization of the issue continues to provide moments of conflict over water infrastructure matters between the two states.

The Historical Roots of Interstate Water Disputes

Contemporary interstate conflicts over water in the postliberalization period in southern India have been shaped in large part by historically produced inequalities and political resentments. A key underlying foundation for such antagonisms can be traced back to the political economy of the colonial state. Contemporary political and economic relationships centered on the sharing of water and water-related infrastructure between Tamil Nadu, Andhra Pradesh, Karnataka, and Kerala are rooted in the geopolitical power of the British-ruled Madras Presidency. As a central site of direct British colonial rule, Madras Presidency pursued its own interests in water resources with the neighboring princely states of Mysore, Hyderabad, Cochin, and Travancore. Unequal relationships between the British colonial state and independent princely states that were heavily influenced by indirect British control allowed the Madras Presidency to develop legal arrangements, irrigation infrastructure, and modes of agricultural development that placed it in an advantageous position over the princely states.

These underlying inequalities of both state power and economic development were incorporated into the new federal structure that would govern relations between the states in postindependence India. Such historical processes are embedded in the dynamics of federalism in the postindependence period. In postindependence India, the formation of Tamil Nadu from the Madras Presidency and the formation of Karnataka, Andhra Pradesh, and Kerala primarily from the princely states of Mysore, Hyderabad, Cochin (Kochi), and Travancore has reproduced colonial political and economic inequalities within independent India's federal structure. These relationships have been embedded in each of the three interstate relationships that the state of Tamil Nadu has been negotiating since the late twentieth century.

While the roots of these political and economic tensions between the water-sharing states of the south can be located in colonial history, they were reworked in distinctive ways through the political dynamics that have shaped the architecture of Indian federalism in the postindependence period. A key element of these dynamics lies in the linguistic reorganization of the states, which both drew on popular social movements and culminated in the States Reorganisation Act of 1956. The reorganization drew the boundaries of the southern states along linguistic lines. For instance, the state of

Andhra Pradesh was formed in response to a popular social movement for a Telugu-speaking state in 1953 and was later expanded to incorporate Telugu-speaking districts of Madras State. Karnataka was formed out of Mysore State and the neighboring Kannada-speaking regions of the Madras Presidency (as well as of the Bombay Presidency and princely state of Hyderabad), and Kerala was formed out of the princely states of Cochin and Travancore, along with a small Malayalam-speaking *taluk* (town) from Madras State. Finally, Tamil Nadu was formed out of the Tamil-speaking Madras Presidency. The conjuncture between this linguistic reorganization and the underlying legacies of colonial relationships of power has meant that the legal and political relationships between the Madras Presidency and the princely states both undergird Tamil Nadu's relationships with its neighboring states and complicate these relationships through ethnicized linguistic cleavages that can become politicized in volatile ways in the context of contemporary disputes over water resources and infrastructure.

In the same historical moment as the linguistic reorganization of the states, the historical formation of the national institutional framework also inadvertently intensified the potential for water conflicts to arise between states. The central government created two sites for the negotiation of interstate relationships over water—the Inter-state River Water Disputes Act, 1956, and the River Boards Act, 1956. However, the River Boards Act was sidetracked by centralized planning in the twentieth century "to develop rivers through interstate planning and development because the Union [central government] controlled the purse strings and the planning process" (D'Souza 2002, 89). The Indian state's institutional architecture was historically oriented toward the mediation and resolution of disputes once they had arisen rather than a policy framework that would promote models of planning and development that would build and strengthen interstate cooperation over water resources.

Contemporary scholarship on interstate water disputes in India has called attention to the deep problems with the institutional mechanisms of adjudication through the central government (Chokkakula 2014; Iyer 2015; Mohan, Routray, and Sashikumar 2010; Moore 2018; Padhiari and Ballabh 2008; Salman 2002; Shah 1994; Swain 1998). Such procedures bring local state governments together in an adversarial judicial framework that makes dispute resolution difficult, hardens polarized positions, and results in lengthy judicial processes that often remain unresolved when tribunal awards are not attached to an adequate institutional capacity or political will for implementation (Padhiari and Ballabh 2008, 189). These problems have led to some institutional reforms of the central government's framework for the resolution of such conflicts. In 2002, amendments to the Inter-state River Water Disputes Act sought to limit the time for the establishment and operation of the tribunal process and gave the tribunal award the same weight as a Supreme Court decision. However, the time frame still remains lengthy, as the amendments allowed for the government to take a year to establish a tribunal, three years with a possible two-year extension for the tribunal to give its decision, and a further year for its report.[2]

These institutional inadequacies were further strained by the developmental policies of the early decades of the postcolonial period. At one level, the absence of effective national regulatory mechanisms was combined with the centralizing authority of the developmental state. In the absence of effective national regulatory mechanisms for the governance of water, the central government's developmental model was implemented without a broader regional or interstate framework for the management of water resources. Such policies have emphasized the expansion of irrigation potential in order to accelerate agricultural productivity and rested on the exhaustive use of water resources through irrigation schemes, groundwater exploitation, and the building of dams to generate electricity and serve the growing need for drinking water supplies. These policies, combined with the historical legacies of the colonial state, produced the extractive structures that have continued to undergird political disputes over water resources.

Both India's period of planned development and its policies of liberalism have rested on an institutional and economic model of development that has produced and exacerbated competition between the states for resources. They have, in effect, created and relied on a political economy of federalism that has weakened interstate and regional cooperation. In the early decades of independence, the central government's planning process emphasized rapid agricultural development and set into place growing strains on water resources for irrigation that are now the foundation for disputes, such as the Cauvery River dispute between Tamil Nadu and Karnataka. State policies in the postliberalization period have intensified rather than broken from this competitive model of state development. In the postliberalization period, interstate competition for both private investment and central government resources that are designed to promote city-based models of development exacerbate the strain on water resources and the corresponding competition between states for these resources.

Economic shifts have intensified pressures on water resources in other ways. For instance, the effects of a shift toward export-oriented models of growth have produced shifts in agricultural production. As economist Narendar Pani has argued, as "agriculture, particularly after the mid-eighties failed to keep pace with the overall growth in a liberalizing economy, food grain production did not always provide the economic returns that the rest of the economy was beginning to enjoy. Farmers could hardly be faulted for moving to some more lucrative non-food crops, even if they were more water intensive" (2010, 51). Meanwhile, the pace of economic growth since the 1990s has also intensified the pressures on water resources, as states must manage a range of demands from industries, farmers, and urban and rural residents.

A comparative analysis of Tamil Nadu's negotiations with its three neighbors illustrates the ways in which such processes are also shaped by patterns of political decentralization that have become a defining feature in India. The shift from a highly centralized polity dominated by the Congress party to a coalition-oriented political landscape in which regional parties and political actors have become significant national players has been a key factor that has shaped the dynamics of decentralization in contemporary India. The consolidation of regional parties in Tamil Nadu and the role of Tamil Nadu in coalitional governments at the national level make it a vital case for analyzing these processes. Such a perspective allows us to move beyond a simplified understanding of decentralization in terms of a shift from power in the center to power in the states. Interstate water negotiations are shaped by the complexities of center-state relationships and national political alliances that affect the interests of both central and state governments.

Institutional Failures, the Political Economy of Interstate Crises, and the Cauvery River Dispute

In September 2016, widespread violent protests broke out in Karnataka and Tamil Nadu over the sharing of water from the Cauvery River. Protests on both sides of the border had been slowly escalating as Tamil Nadu began pressing for the release of water according to the final judgment of the

Cauvery Tribunal in 2007 and Karnataka continued to appeal the judgment through the Supreme Court. Farmers on both sides of the border were facing water shortages, and Bengaluru (Bangalore), India's high-tech center, was facing drinking water supply shortages. When Karnataka was eventually compelled to abide by an interim Supreme Court decision to release 13.6 TMC of water to Tamil Nadu over a period of ten days, protests in Karnataka intensified. Bus services and other traffic between the states were canceled, heavy security had to be deployed at reservoirs of the Cauvery Basin, and protests escalated into ethnic violence directed against vehicles and hotels.[3] Violence in Bengaluru was particularly bad, with the unrest effectively shutting down a city that has been branded as the face of the success of a liberalizing India. In the immediacy of the conflict, the combined effects of media sensationalism (including both regional and English-language news and social media reports), rife with language about "water wars" and the politicization of local ethnic-linguistic identities, deepened the polarization between the two states.[4] The politicization of the dispute through political party competition both at the local state level and in terms of national politics played an important role in deepening the conflicts.

The deep-seated sources of such conflicts lie in the entangled roots of state water policies of the colonial, twentieth-century developmental, and twenty-first-century liberalizing state. By the beginning of the twentieth century, the dispute over the sharing of the Cauvery River between Karnataka and Tamil Nadu had already become one of the most infamous examples of the inability of the Indian state to effectively mediate conflicts over water resources. As with many of India's water-sharing disputes, the initial roots of the conflict can be traced back to colonial legal and political agreements that set up a hierarchical political and developmental relationship between the Madras Presidency and the princely state of Mysore. This resulted in a series of legal agreements and political negotiations designed to protect British colonial interests in agricultural development in the Madras Presidency by placing limits on Mysore's ability to construct new irrigation projects. Mysore, in effect, needed the consent of the British colonial state in order to engage in new projects that would potentially affect water supplies to the Madras Presidency (Benjamin 1971).

The unequal political-economic structures of agricultural development that emerged from this colonial history began to change rapidly in the early decades of independence. Irrigation along the Karnataka side of the Cauvery

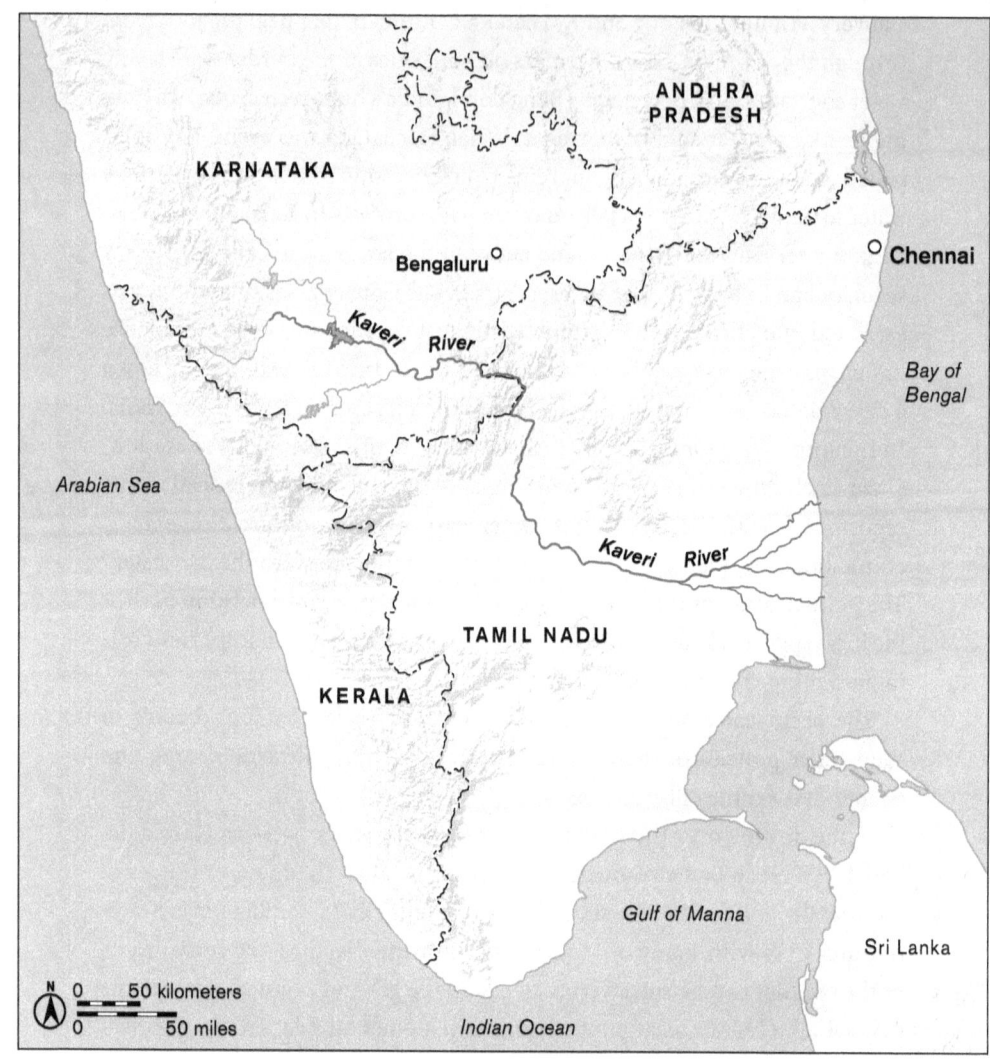

MAP 3.2. The Cauvery River, depicting the Kaveri (Cauvery) River, which has been the site of India's longest river-sharing dispute, between Karnataka and Tamil Nadu

did not, for instance, begin to accelerate until the 1960s (Guhan 1993, 6). The area of Cauvery irrigation in Karnataka increased from 442,000 acres in 1971 to 2,138,000 gross acres in 1990, with a corresponding increase in water utilization requirements from 110.2 TMC to 322.8 TMC. Meanwhile, Tamil Nadu's irrigation increased marginally, from 2,530,000 gross lakh acres in 1971 to 2,580,000 in 1990, with a corresponding increase in water utilization requirements from 494.6 TMC in 1971 to 501.5 TMC in 1990 (Guhan 1993, 21). This shift meant that Tamil Nadu could no longer rely on a reliable release of water from the Cauvery River. Tamil Nadu's historical advantage was significantly reversed by its geographical location downstream from the river. The political-economic context that provided the need for a resharing of the river had been laid.

The overexploitation of the river was intensified by the Indian state's ambitious centrally directed developmental agenda, designed to engage in the accelerated expansion of both industrial and agricultural production. The harnessing of river waters through large dams and diversion canals was central to India's planned economy, with rapid increases in state investment in the early decades of independence. As Ashok Swain has noted, in "1948, 160 large water projects were being considered, investigated or executed, and 2 years later 29% of the first five year plan (1951–55) budget was allocated for this purpose.... Before the eighth plan, 600 billion rupees had been spent for various major and medium irrigation projects" (1998, 168). Budgetary allocations provided important incentives for state governments to embark on strategies of agricultural growth that would expand the exploitation of water resources for irrigation purposes. This centralized framework of planning did not incorporate within it any focus on regional development that could potentially provide an institutional or economic foundation for cooperation between states. The River Boards Act, 1956, was never integrated within the water-intensive planning model of agricultural development. The result in the case of the Cauvery Basin was that Karnataka engaged in the rapid development of irrigation systems, including the construction of a series of dams that heavily reduced water available for Tamil Nadu (Swain 1998, 173). The developmental imbalance that had been produced in the colonial period was rapidly reversed in the early decades of independence. The distinctive nature of this conflict has subsequently rested not simply with a question of sharing river resources but with the task of "re-sharing a heavily used river, involving difficult adjustments" (Iyer 2003, 2350). In distress years, farmers from both states suffer deep consequences from the lack of water resources.[5] Food production and the subsistence of millions of people in both states depend on water from the river (Janakarajan 2010). This overexploitation of the Cauvery River set into motion the decades-long conflict between Karnataka and Tamil Nadu.

In recent years, state-led policies of liberalization have continued to exacerbate these pressures on the Cauvery River, as rapid urbanization and city-based models of economic development that undergird such policies have intensified demands for water resources for both urban drinking water supplies and industrial sources. In the case of the Cauvery River, the city of Bengaluru (Bangalore), which is often branded as the IT capital of the country and serves as one of the most visible symbols of India's economic growth, relies on water from the river as one of its key sources of drinking water supply. While institutional, political, and media narratives focus on the intensity of the conflicts between the states of Karnataka and Tamil Nadu, pressures on water resources caused by economic growth and urbanization have deepened the desperation over claims on the river water in more nuanced ways.

The lens of interstate conflict often blurs the ways in which such water-related stress is as much about inequalities and conflicts within the states in question as it is about conflicts between states. In Karnataka, drinking water resources from the Cauvery Basin are channeled to Bengaluru at the expense of smaller urban and rural localities (Saldhana and Rao 2015, 301).[6] As Leo Saldhana and Bhargavi Rao have shown, "Farmers from Mandya and Mysore in Karnataka, who have vehemently objected to the release of Cauvery waters to downstream Tamil Nadu during droughts, have begun targeting the supply of water to Bangalore [Bengaluru] in protest" (297). This has been institutionally embodied in the growing centralized authority of Bengaluru's municipal water utility, the Bangalore Water Supply and Sewage Board. The supply of drinking water from the Cauvery River to Bengaluru has been structured through this "highly extractive, centralized and financially indebted enterprise" (Goldman and Narayan 2019, 102). Such dynamics point to the hidden complexities of local state authority that lie beneath what seem like intractable interstate conflicts. Or, to take another instance, tensions between Karnataka and Tamil Nadu over the release of water are also shaped by the drinking water needs of Bengaluru during periods of water scarcity. As Karnataka's water minister would point out, the drinking water requirements could not be met if the state government released water from the Cauvery reservoirs for crops.[7] Such examples illustrate that the underlying structural conditions of the Cauvery dispute are as much about inequalities and competing demands within states as they are about the subsequent intransigence of competing local state governments.

These competing demands for water are intensified by a skewed institutional framework produced by policies of economic liberalization. While in the era of planned development local state governments competed for resources from central government budgetary allocations, they now compete for private capital. This competition has been encouraged by the central government without any institutional framework that has simultaneously promoted cooperative regional or interstate models of growth or development. This centralized framework of reforms that undergirds local state governmental economic policies is marked by an absence of institutional mechanisms that can govern relationships between competing states.

The absence of an interstate institutional framework of economic cooperation has meant that interstate relations are managed through a centralized institutional framework only once they have reached a point of polarization in the form of a dispute that must be adjudicated either through a central government tribunal or by the Supreme Court. The turn to these centralized mechanisms for dispute resolution is an effect of the simultaneous absence of effective regulatory mechanisms on the one hand and the centralized framework of economic development and growth that has promoted interstate competition (which in turn has intensified the local centralized power of state governments) on the other hand.

The weakness of regulatory mechanisms for interstate cooperation over water sharing produces new strains on the state, which unfold through dissonances within the institutions of the central government and can then erupt in political conflicts. The result is a volatile set of negotiations and relations between local state governmental and civil society actors within the states and central institutions such as the Supreme Court, tribunal committees, and the central government itself. The interstate dispute over the Cauvery River reveals ways in which the incapacities of the central government—in this case in the form of the absence of an effective regulatory framework—in turn reinforce more subtle and variegated forms of centralization at both central and local state governmental levels.

Consider the institutional dynamics of the Cauvery Tribunal. The tribunal delivered its final judgment after seventeen years of adjudication, from 1990 to 2007—a period that does not include twenty-six ministerial meetings and negotiations that took place in the preceding twenty-two years (Richards and Singh 2002). The tribunal award was finalized only in 2013, after a Supreme Court intervention directing the central government to notify the award so that it could come into force. Furthermore, the central government continued to stall on the establishment of the regulatory institution, the Cauvery Management Board. According to the terms of the tribunal award, the central government was instructed to set up both the Cauvery Management Board and the Cauvery Water Regulation Committee in order to provide the institutional mechanisms to manage implementation of the agreement.[8] However, the award was met by continued central government inaction.[9]

A primary cause of the paralysis of the central government rests with the political strains on action on behalf of either state. While the rise of regional parties and coalition governments has often been characterized as a positive feature of the deepening of federalism in India, the nationalization of local political dynamics can also place significant strains on effective federal governance. In the case of the Cauvery dispute, the federalization of politics has weakened the potential for cooperative federalism. Consider the ambivalent role of the Modi-led government that came to power in 2014. On the one hand, as a member of the ruling coalition, Tamil Nadu's AIADMK-led government had political weight with the Modi administration. On the other hand, the BJP's attempt to regain power in the state of Karnataka produced counterpressures on its actions. In the run-up to the 2018 state elections, the Modi government specifically delayed the establishment of the Cauvery Management Board (a term of the tribunal award that Karnataka has been opposed to) because of the elections. The reason the central government provided to the Supreme Court, that "the PM and ministers are busy in Karnataka polls and can't approve the scheme for releasing water to Tamil Nadu, as directed," was a thinly disguised sign of the political risks of setting up the board prior to the elections.[10] Meanwhile, opposition political parties in Tamil Nadu kept up continual political pressure on the AIADMK-led government.

The combined result of such central government paralysis and the pressures on water resources that have been intensified by layered sets of state economic policies has meant that water sharing between Tamil Nadu and

Karnataka has continued to exacerbate tensions between the states. The combination of real water distress produced by failed monsoons (for instance in 2012–13 and in 2016–17) and intense political opposition to the terms of the tribunal award has led Karnataka to refuse to release water according to the terms of the tribunal judgment. Tamil Nadu, on the other hand, faced with the severity of its own water scarcity with the failed monsoons and with the challenges of managing its ongoing water insecurity, has aggressively turned to the courts to enforce the agreements.

In response to a Tamil Nadu petition to the Supreme Court to mandate the formation of the Cauvery Management Board by the Ministry of Water, the Supreme Court directed the central government to set up an interim panel to arrange the release of Cauvery waters in 2013. However, the terms of water sharing between the two states remained unresolved and politically volatile, with fierce conflicts breaking out in the context of the interim committee.[11] Once again, on September 20, 2016, central government inaction prompted the Supreme Court to intervene and direct the central government to set up the Cauvery Management Board within four weeks. However, the central government refused to implement the order, arguing that the court did not have jurisdiction over the matter.[12] As the attorney for the central government argued,

> Since the central government was not a party to the proceedings before the Tribunal, it did not have opportunity to submit to the Tribunal that the Tribunal will not have any power of recommending to create a Board as suggested. Setting up of a Board is part of legislative exercise. . . .
>
> It is submitted that constitution of a Board as suggested by the Tribunal and ordered by this Court on September 20 is not contemplated by the statute. By setting up of a Board of this nature, the Central Government is denuded of its power under the Act of 1956 to frame a scheme based on an award which goes through a legislative process by placing thereof before the Parliament and the final say is vested in the Parliament.[13]

The state, in effect, argued that the tribunal did not have the authority to recommend the constitution of a water board without the consent of Parliament. The central government was in effect trying to draw boundaries around the power of the Inter-state River Water Disputes Act, which had been amended precisely to avoid lengthy delays in implementing tribunal awards. The absence of central government action once again prompted the Supreme Court to assert its own authority and compel the central government to form the Cauvery Water Management Board (which it did after the Karnataka elections).[14] Such interventions of the Supreme Court in the years after the tribunal award in themselves reflect a failure of the central government's regulatory state capacity.

According to a 2002 amendment to the Inter-state River Water Disputes Act, the tribunal award was to be "final and binding on the parties to the dispute" (PRI 2002, 4). The amended act specifically sought to circumvent potential lengthy (and politically volatile) challenges through the Supreme Court by giving the judgment the binding authority of a Supreme Court decision. According to the act, "The decision of the Tribunal, after its publication in the Official Gazette by the Central Government under sub-section 1, shall have the same force as an order or decree of the Supreme Court" (4). In light of this amendment, the fact that the Supreme Court had to intervene and direct the central government to notify the award after six years of inaction was a vivid sign of the central government's institutional and political failure in managing the dispute.

What we see unfolding is a deeper dynamic of state incapacity, as the continued inability of the central government to manage the implementation of the award produced an institutional vacuum that compelled the Supreme Court to step in. Yet the 2002 amendment to the Inter-state River Water Disputes Act, 1956, had sought to prevent precisely this kind of situation, where the Supreme Court had to take on an expanded role in order to respond to the paralysis of the central government's executive authority. Evidence of this dynamic has been highlighted by a new amendment to the interstate waters dispute act (LS 2019), which has passed the Lok Sabha and is designed to address these ongoing problems.[15]

The dynamic surrounding the Cauvery River dispute provides a historic case of the institutional incapacity of the Indian state that has shaped the management of the political ramifications of the very real economic distress produced by periods of water scarcity. Meanwhile, the intervention of the Supreme Court in the vacuum provided by this form of state incapacity has complicated the dispute resolution process and produced the very kinds of judicial appeals and claims that the amended Inter-state River Water Disputes Act was intended to avoid.[16] Supreme Court judgments in the case in the period between 2013 and 2016 in fact reflect the actions of a court attempting to force the government to take executive action—for instance by notifying and implementing the award of a tribunal that the central government's own legislative action had sought to define as final, binding, and with the "same force" as an order of the Supreme Court. Furthermore, in the context of dire water scarcity (during periods of drought) within both Tamil Nadu (without the release of sufficient water since the notification of the award) and Karnataka, the court has attempted to manage an emergency situation that should have been the responsibility of the central government. In this case, the scarcity of national regulatory state action in the Cauvery case has intensified the consequences of water scarcity in both states, which have been exacerbated by increasing demands on water in the context of state-led policies of liberalization.

The result of this state incapacity has been that both competing central institutions such as the Supreme Court and local state governments have filled these institutional gaps. This has deepened the polarization between the states on matters related to water resources and hardened the desire of state governments to assert control over water resources and infrastructure that in any way impacts such resources. This is evident in the ways in which developmental water-related infrastructure has increasingly become a kind of weapon that continues to exacerbate political tensions over the Cauvery River. Tamil Nadu has sought the central government's intervention to prevent Karnataka from building infrastructure that would impact the use of the river's resources, for instance by protesting Karnataka's proposal to build a dam and reservoir for the generation of hydroelectric power and provision of drinking water for Bengaluru.[17] Meanwhile, Tamil Nadu has itself sought to build a large dam across the river (Saldhana and Rao 2015). Water-related infrastructure becomes a means for political mobilization within the contours of normative state visions of economic development and growth that intensify the very inequalities and forms of scarcity that deepen the distress for local communities in both states. The result is a subtle but significant intensification of local state governmental authority over water. Furthermore, the politicization of water conflicts has led to increased state governmental competition for central government intervention on behalf of their interests.

Consider the ways in which the Cauvery dispute was intensified by the politicized nature of interstate competition in the decades-long tribunal process. At an early stage of the process, tribunal members toured the Cauvery Basin on both the Karnataka and the Tamil Nadu sides of the river. Both states felt the political pressure to impress the tribunal members not just through the technical and legal arguments emphasized in the formal proceedings but through the social and political rituals of the tour. Tamil Nadu's head of the Cauvery technical cell of Karnataka's weeklong tour, held in 1991, described it thus: "One weeklong tour in Karnataka Cauvery Basin, along with the Chairman and Members of the Cauvery Waters Disputes Tribunal, in a big convoy of 40 cars and saw rousing reception wherever we went. Tea and extraordinary lunches and dinners all through were arranged by the Karnataka Cauvery team. . . . The Karnataka State Government has flexed their muscle to make this Tribunal tour a memorable one for their own benefit" (Mohanakrishnan 2016b, 31).

This narrative does not imply that such social processes influenced the members of the tribunal. Rather, what matters is that a more textured perspective on the role of state governments illustrates that they are operating within an institutional framework which is itself embedded in political processes. While politicians indeed further politicize issues through public and political attacks on opposing parties, such examples reveal the ways in which the institutional mechanisms of the dispute reconciliation process themselves become immersed in political processes that are shaped by the hardened and polarized state governmental authority of each of the states involved.

Consider, for instance, some of the more public political performative dimensions of the tour that Tamil Nadu in turn organized for the tribunal members. Tamil Nadu organized a three-hundred-kilometer-long human chain along the route that the tribunal members' convoy traveled during the tour along with farmers' meetings and a final rally (Mohanakrishnan 2016b, 33). The organization of this human chain provides insight into the complex nature of state-society relations in the context of the dispute. The formation of the human chain was a product of both state-led and farmers' political organizing. It would be a mistake to conceive of farmer mobilization purely as a state-driven process. On the contrary, real problems of water scarcity and economic crises (in the context of failed crops during distress years) have meant that farmers in the state have actively mobilized and often taken the lead in placing political pressure on the state government to address diminishing waters from the Cauvery. The Cauvery Tribunal itself was formed by a Supreme Court directive in response to a petition filed by the Tamil

Nadu Farmers Society (Swain 1998, 173). Nevertheless, such protests within civil society have facilitated the exercise of state governmental power in the dispute. For instance, the Public Works Department, working with local collectors in the districts, played a central role in organizing farmers. The state both foregrounded the significance of the tour and provided institutional mechanisms that helped facilitate the popular response from farmers. State–civil society relations unfolded in ways that consolidated local state power rather than in ways that deepened or expanded the space for democratic political participation.

The dynamics around the Cauvery Tribunal since 1990 have taken place in a political context in which there has been a shift away from Congress party dominance to a more complex national pattern of coalitional politics. The unwillingness and inability of the central government to effectively intervene in the management of the dispute has spanned both Congress and BJP-led governments, as neither party has had the political will to weaken their political influence within the states. Meanwhile, the dispute over the river has become a significant site for political mobilization within both Tamil Nadu and Karnataka, with the various political parties either accusing their opponents of being weak on the issue or risking being the subject of such accusations. The potency of political mobilization has been accentuated both by the cultural significance of the river for communities (Settar 2010) in both states and by the potential politicization of ethnic-linguistic differences between the states (Pani 2010). Such processes of politicization have been hardened in ways that have facilitated the exercise of state power within civil society while polarizing relations between communities of the two states. Protests and demands by civil society actors, in this context, tend to intensify the pressures on state governments to claim sovereign authority over water.

The Cauvery dispute is illustrative of a case in which state–civil society relationships in both states have been polarized by the cumulative effects of decades of state institutional incapacity. As the comparative framework of this chapter illustrates, there have been significant variations in Tamil Nadu's negotiations over water-related matters with its neighbors. The kind of water resharing required under stringent political-economic conditions and the political paralysis of the prolonged adjudication makes the Cauvery dispute a unique situation. Yet its very distinctiveness provides an illustrative case for an understanding of how the institutional incapacity of the state—in this case the weakness of regulatory mechanisms—has meant that the local state governments have hardened the state–civil society compact within each state in ways that have deepened conflicts between Karnataka and Tamil Nadu.

One of the most significant, though rare, attempts at redrawing these state–civil society relationships by attempting to build civil society relationships between Karnataka and Tamil Nadu was the Cauvery Family initiative. The Cauvery Family initiative took place between 2002 and 2013 and sought to build civil society linkages and dialogues between farmers in Karnataka and Tamil Nadu and to include an array of "academics, bureaucrats, NGOs, lawyers, people from the media, and other concerned citizens" (Janakarajan 2010, 150). Professor S. Janakarajan, founder of the initiative, wanted to build what he has called a multistakeholder dialogue, which would in effect develop a set of both civil society and institutional linkages across the territorial boundaries of the states. The initiative has been held up as a unique example of an alternative approach to the protracted conflict. Years after the conclusion of the initiative and in the midst of the severe violence that unfolded in 2016, Janakarajan would write, "The initiative, Cauvery Family, has met eighteen times. In our last meeting in 2012, we arrived at a water sharing formula acceptable to farmers in both states. Though this initiative was widely appreciated by the media and civil society, it failed to grab politicians and governments' attention. The initiative failed as it did not receive any political support. The violence we are witnessing in the two states could have been circumvented had the political parties or governments recognized initiatives by non-governmental/non-political organisations."[18]

At one level, Janakarajan's reflections on the lack of responsiveness of both governments and politicians points to the constraining effects of both the state's institutional framework and the deep regulatory weaknesses in the governance of water. The Cauvery Family initiative was in effect trying to build the very institutional mechanisms for reconciliation that have been missing in the existing framework for water resources management that became entrenched in postcolonial India. At a deeper level, they point to the ways in which this prolonged state failure also produces a particular state-society configuration that has intensified the ethnic, linguistic, and territorial divisions between the two states during periods of water scarcity and that has in turn reconsolidated state governmental claims of authority over water.

The enhancement of sovereign claims of state governmental authority in the context of heightened competition between Karnataka and Tamil Nadu has also produced subtle foreclosures in the space for effective bureaucratic agency and negotiation. Consider, for instance, the internal dynamics of the Cauvery Technical Cell documented by the chairman. Mohanakrishnan's professional records provide a unique view of the daily work of negotiations, preparation and presentation of technical data, and filings of court briefings, often at a moment's notice at the request of the Supreme Court or state governmental leaders. The records show Mohanakrishnan's keen critical views of time spent in unproductive bilateral and intrastate meetings (2016b, 24–36) and the waste of resources in the protracted legal proceedings. In one entry on the proceedings of the tribunal, he observes that there were, "of course, a few undeserved Advocates who did nothing, who spoke not a word in the court, who simply sat in the Advocates' conference and the courts and drew their fees in lakhs, we could see, but that is beside the point. They got themselves included in the team on political influence" (44).

Weary as he gets when there is "more of gossiping than serious work" (96), Mohanakrishnan's concerns about waste are not limited to discussions of lawyers but punctuate his records on the multiple meetings, conferences, and events organized for and around the Cauvery dispute. They point to a little-discussed dimension of state practices in scholarly work that has tended to be highly critical of the role of state bureaucrats, professional expertise, and governmental corruption; that is the significance of understanding spaces of ethical agency that exist within the state. Thus, for instance, he characterized his appointment as head of the Cauvery Technical Cell as the start of "the decades-long journey of seeking justice" (2016b, 26). To fully understand this ideal of justice, this characterization must be distinguished from the visible dramatic and public rhetoric of political leaders that invoke norms of justice with broader public or electoral calculations at hand. Indeed, at points in his personal journal, Mohanakrishnan notes (in an understated tone) the pressure of needing to save crops in the Delta region of Tamil Nadu that lies at the backdrop of the endless negotiations, meetings, and travel that he is engaged in (Mohanakrishnan 2016b).

The potential for understanding such spaces of bureaucratic agency within the state provides an avenue for moving past static narratives of state dysfunction that have become a key dimension of public and political rhetoric in the postliberalization period. Such a textured conception of state agency can also begin to move us beyond rigid views of inevitable deadlocked forms of political polarization and conflict over water resources. One of Mohanakrishnan's overriding concerns, for instance, is with the accuracy of the technical data being presented. Consider, for example, the perspective of one of Karnataka's eminent lawyers who represented the state in the Cauvery dispute: "My own experience in the Cauvery Water Disputes Tribunal has been if the Chief Engineers of Karnataka, Kerala and Tamil Nadu had been assembled to sit across the table with the Chairman (and members) of the Tribunal, it would have been possible to narrow differences and save a great deal of time. . . . The engineers had to be put at ease so that they did not have to keep looking over their shoulders (to their masters, the State) when explaining technical matters" (Nariman 2009, 52).

This potential for a constructive, nonpoliticized process of conciliation, of course, as we have seen, was foreclosed by both state executive inertness and political polarization at the local state and central governmental levels. The result, as S. Guhan has argued, was that "expert engineers on both sides were not able to quietly work together to find common ground; on the contrary, they got co-opted to advance or defend partisan positions" (1993, 35). While state and political fractures overwhelmed the conciliation process, such perspectives point to the significance of taking seriously the role and potential of actors within state bureaucracies who may open up or obstruct the spaces for the effective reconciliation of disputes and expand the space for interstate cooperation.

This range of bureaucratic activity lies in a liminal space between the visible drama and rhetoric of political leaders in the central offices of state governments on the one hand and the inert cultures of the bureaucratic state on the other. It is this kind of hidden work of bureaucratic actors that is perhaps the least analyzed dimension of current social science research on contemporary India. Yet local state actors are themselves negotiating within the institutional constraints and incapacities and political-economic structural conditions. They, in effect, perform the everyday labor of the state. It is the measured observation and analysis of an intermediary local state official engaging in the arduous daily actions of pressing the case forward to a kind of resolution that is the substance of the kinds of practices that institutional reforms seek to put into place. An understanding that disentangles bureaucratic action from the centralized nature of state action at both the local and central governmental levels is thus crucial for an adequate analysis of the dynamics of institutional reform.

Interstate Cooperation, the Telugu Ganga/Krishna Water Supply Project, and the Spaces of Bureaucratic Agency

If the Cauvery River dispute has become an infamous example of institutional failure, the Telugu Ganga Project that produced an agreement between Tamil Nadu and Andhra Pradesh is often heralded as a model of interstate cooperation (Sampathkumar 2005). The agreement centers around the Telugu Ganga Project (formally known as the Krishna Water Supply Project), which supplies water from the Krishna River for Chennai's drinking water supply and for irrigation needs in Rayalaseema, a drought-prone area in Andhra Pradesh. The structural conditions and substantive focus of the agreement are fundamentally different from the Cauvery case. Historically, Andhra Pradesh was a part of the Madras Presidency, and the state was first carved out of the Telugu-speaking areas of Madras State in 1953 (with Telugu-speaking areas of Hyderabad State joining the state as part of the state-reorganization process in 1956). While there is a history of politicized linguistic distinctions between Tamil Nadu and Andhra Pradesh, the two states are not shaped by a colonial history of sharp disputes, which has characterized the Tamil Nadu–Karnataka relationship. Bureaucratic officials in Chennai also argue that the two states have shared strong cultural and economic ties because of the links between members of the state bureaucracy in Andhra Pradesh and the city of Chennai (interview, PWD, January 2017). Such ties were accentuated during the early years of negotiation between the two states by the shared background of two chief ministers, M. G. Ramachandran (MGR) of Tamil Nadu and N. T. Rama Rao (NTR) of Andhra Pradesh. Both chief ministers, representing independent regional parties, came to politics as highly successful stars in regional films. NTR, in particular, also had ties to the Tamil film industry. Such ties were a highly visible example of deeper ties between Andhra Pradesh and Chennai-based cultural and economic activity.

A second crucial difference between the Cauvery dispute and the Telugu Ganga agreement lies in the nature of the cooperative water sharing that was institutionalized. The agreement, formally signed in 1983, represented cooperation over a set of shared interests. The Krishna River, which provides water resources through the states, does not run through Tamil Nadu. Disputes over the sharing of the river water have played out through a separate Krishna Water Tribunal, which has mediated conflicts between the states of Andhra Pradesh, Karnataka, and Maharashtra.[19] The Telugu Ganga agreement served the mutual interests of both states, as it allowed Andhra Pradesh to supply water for a drought-prone area in addition to providing water for Tamil Nadu.

Finally, the political conditions of the central and state governments also played a significant role in jump-starting the agreement. The agreement grew out of a Congress government–led initiative that gained the consent of the states of Maharashtra, Karnataka, and Andhra Pradesh to each provide five TMC of the fifteen TMC to Tamil Nadu in 1976. This would later become the basis for the bilateral agreement providing fifteen TMC from a reservoir in Andhra Pradesh for Chennai's drinking water (Mohanakrishnan 2011a 12). The cooperation between the states was in effect a product of Congress's one-party rule in the early decades of independence as well as Indira Gandhi's questionable use of executive authority. The agreement was executed in the context of Indira Gandhi's suspension of democratic rights during the Emergency period. More specifically, in the context of Tamil Nadu, Gandhi had dismissed the elected government on February 15, 1976, and instituted president's rule. The announcement of the agreement for Chennai's water supply was part of a visible political ritual that Gandhi was using to produce consent to her political actions. For instance, she visited Madras (now known as Chennai) two weeks after instituting president's rule to announce the agreement. The publicity around the project was of course a strategy designed to gain popular support within Tamil Nadu in the context of Gandhi's antidemocratic actions both at the national level and in Tamil Nadu. Gandhi's authoritarian actions during the Emergency had garnered support from significant sections of the urban middle classes. Moreover, the inauguration of infrastructure projects has long been a strategy that elected officials have used to gain electoral or popular support (Min 2015). Gandhi was, in effect, using the promise of drinking water through a large infrastructure project for the city of Chennai as a political strategy to garner public support in the face of her government's political ousting of the elected state government.

If, in the Cauvery River dispute, complications associated with the fragmentation of the political field produced a kind of state paralysis, paradoxically, one-party rule provided an important catalyst for the initiation of the Telugu Ganga Project. Seven years later, when the official agreement was signed between the chief ministers of Tamil Nadu and Andhra Pradesh, Indira Gandhi, in power once again, could attend the public ceremony and hold up the project as an emblem of national unity, noting that "it has long been a dream of mine that the rivers of India should be joined together, joining the different peoples, the different cultures together. Bringing water to parched lands and also opening out new ways of transport" (quoted in Mohanakrishnan 2016b, 93). The infrastructure project allowed her to link "the common man," states that are "strong and self-reliant," and the "strength" of the central government through a narrative of national unity. As she put it, "We should all regard ourselves not merely as citizens of Tamil Nadu or Andhra Pradesh or Maharashtra or Karnataka or anywhere else but as citizens of India bound together in a comman [sic] goal of making our country self-reliant, strong, unified and great. We want strength, not to dominate over anybody, or any other country, any other people. But to be able to protect ourselves and solve our multifarious problems" (94). The rhetoric captured the kind of centralized federal structure that was at the heart of Gandhi's approach.

The contrast between the central government's interventionist action in the Telugu Ganga case and the incapacitated nature of state action in the Cauvery case would seem to provide an argument in favor of centralized state action. However, while state action in the Telugu Ganga case did provide a critical factor that set up the foundation for the project, similar action did not prove to be effective in the Cauvery River dispute during the same time frame. In contrast to the Telugu Ganga case, in the Cauvery case, the political and institutional turmoil of the Emergency period foreclosed an agreement that was being forged between Tamil Nadu and Karnataka.[20] In 1971, a coalition of farmers with support of members of both regional and Congress parties in the two states sought an agreement via the Supreme Court. In that context, Gandhi once again intervened and produced a draft agreement in 1976 that had consensus from the two states.[21] However, during the Emergency, the farmers' case was dismissed by the Supreme Court, and in light of the deterioration in center-state relations, the newly elected DMK-led regional government in Tamil Nadu refused to support an agreement developed through Gandhi's intervention. Centralized state intervention and the turmoil produced by the Emergency in the Cauvery case broke down the 1976 consensus and contributed to the acceleration of the dispute into an irreconcilable conflict that was referred to the tribunal in 1990.

The confluence of a set of shared central and state governmental interests does not fully account for the divergent paths in the Cauvery and Telugu Ganga cases. The deep developmental strains on the Cauvery Basin and the shared strategic interests between Andhra Pradesh and Tamil Nadu are significant factors that explain this divergence. However, a careful analysis of the process and politics surrounding the Telugu Ganga infrastructure project reveals a messier process that is not adequately captured by an explanation of shared strategic interests. Consider, for instance, the passage of time between the initial agreement in 1976 and the completion of the project. It took seven years for the signing of the bilateral agreement to take place in 1983; the public inauguration marking the implementation of the project was held after an additional thirteen years. Since then, the supply of water to Chennai has been uneven. While the formal bilateral 1983 agreement has often been heralded as a model of interstate cooperation, the successful implementation was not an inevitable outcome. Institutional records of the Public Works Department in fact show a much more complex process that moved forward in the face of continued political obstacles. The completion of the project was in large part due to effective technical and bureaucratic agency at the local level. Moreover, while water supplies still do not flow reliably to Chennai, the long-term impact of the prolonged but effective institutional cooperation at the local level has meant that the political and institutional space for negotiation has remained open, in contrast to the Cauvery dispute. While shared political and structural interests were necessary conditions for the establishment of this form of interstate cooperation, they were not sufficient for the implementation of the project.

Institutional records of the Public Works Department reveal numerous delays and roadblocks that could have potentially derailed the successful implementation of the agreement. Such implementation—which entailed the physical creation of the infrastructure needed for the water supply—was successful because of prolonged and persistent cooperative efforts at the local level. A key foundation for this cooperation was the creation of microinstitutional mechanisms for communication and cooperation between state actors and technical experts from the two states. For instance, two committees, a Liaison Committee composed of state officials and a committee of technical officers, provided crucial means for communication, which allowed for the management of the project in ways that circumvented the escalation of differences into wider political battles (Mohanakrishnan 2011a 20–22).

The interstate agreement is often heralded as a model because it served the interests of both Tamil Nadu and Andhra Pradesh. However, this establishment of shared interests was formulated after the initial agreement was ratified in 1977. The 1977 agreement did not contain any provisions for supplying irrigation waters for Andhra Pradesh, and this demand was not made by Andhra Pradesh until the fourth meeting of the Liaison Committee, in 1979 (Mohanakrishnan 2011a, 22). The demand was then reiterated through a specific proposal to irrigate Andhra Pradesh's drought-prone area made in 1980. While this would become part of the final agreement, records show that key local decisions produced a pragmatic solution to what could in a more polarized political context have escalated into an obstacle to, if not breakdown of, the project. A. Mohanakrishnan, Tamil Nadu's chairman of the Committee of Technical Officers and member of the Liaison Committee, would later note, "This [demand of Andhra Pradesh] put the State of Tamil Nadu in a somewhat piquant situation requiring careful decisions to be taken after deep thinking" (24). Mohanakrishnan's solution was to develop an idea of a combined conveyor system from the reservoir source (Srisailam) located in Andhra Pradesh (see map 3.3). This technical solution provided a safeguard for Tamil Nadu, as it would serve as an insurance that any future problems with the canal would not leave the state in a powerless situation. In other words, a breach in Andhra Pradesh's canal, which would have been bigger and at a higher level, would have potentially overwhelmed the supply channel to Chennai. The proposal for the joint canal accommodated Andhra Pradesh's needs but in a way that safeguarded some of Tamil Nadu's interests in the future operation of the system (25). While Andhra Pradesh initially held back on this proposal, patient negotiations within the committee structure eventually produced a foundation for the actual implementation of the agreement.

Such processes reveal the intricate negotiations through an institutional form as mundane as a joint committee structure that can too easily be glossed over by simply focusing on the formal signed agreements based on shared economic interests. Consider, for instance, another key obstacle to the implementation of the agreement—the question of funding. The construction of a joint canal meant that the two states had to come to a cost-sharing agreement for the project. While such an agreement was reached, the question

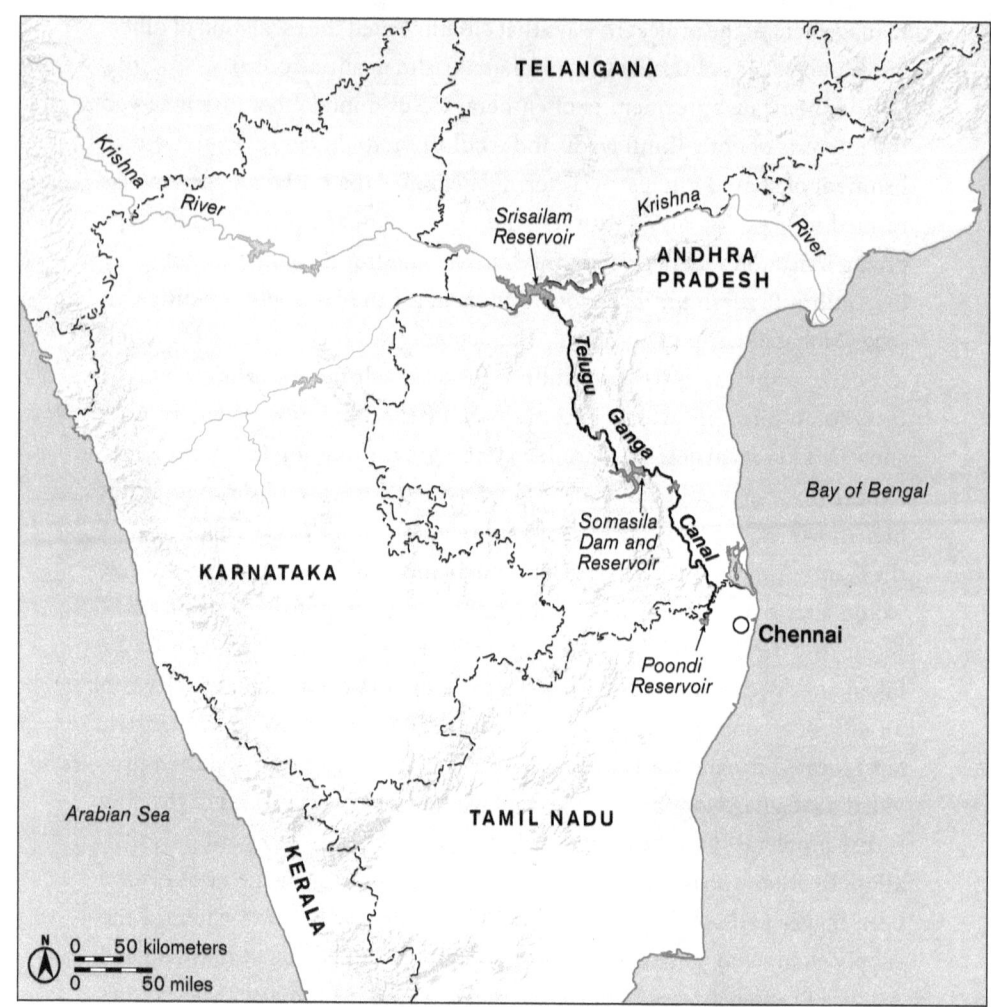

MAP 3.3. The Telugu Ganga/Krishna Water Supply Project, illustrating the route of the water supply infrastructure project designed to supply water from Andhra Pradesh to the city of Chennai

of who would front the funds in the initial stages of the project became a source of contention between the states. As early as 1986, the chief minister of Andhra Pradesh wrote to MGR, his counterpart in Tamil Nadu, expressing concern about inadequate funds released for the project and requesting a meeting between the two chief ministers (cited in Mohanakrishnan 2016b, 116). MGR replied to this concern with an ambivalent response, indicating that "it is desirable that the officials of both the States meet once again and discuss the status of the Project and other related issues, before we meet for a discussion" (117).

While a meeting between the two chief ministers was eventually scheduled, it led to a breakdown in the implementation of the project. Mohanakrishnan described the events in the following way:

> We were all waiting, both the Ministers and Officials groups from both the Governments, in the Secretariat from forenoon. But the meeting did not take place since a message was received that the Hon'ble Chief Minister of Tamil Nadu had suddenly fallen sick. The whole Andhra group returned in the evening. On his way to the airport, Sri N.T. Rama Rao called on his brother Chief Minister of Tamil Nadu reported sick, presumably to enquire about his health. There was one-to-one meeting. No one knows what transpired in that meeting. But the flow of funds for the project stopped thereafter. It was true that the work on the project had a bad set-back for nearly two years. (2016b, 63)

The weighty silence embedded in the words "No one knows what transpired in that meeting" is a vivid reflection of the political fragility of interstate relationships. The issue at stake was not the terms of cost sharing but the question of which state would provide more of the funds upfront. What Mohanakrishnan's records reveal are increasing concerns within the Tamil Nadu state government about the high costs of the project (Mohanakrishnan 2016b, 61). It would take a change of government in 1989 for Tamil Nadu to commit funds to the project. Such funding disputes would continue to affect the implementation of the project, particularly as delays would mean a continued increase in the costs of the project.[22] Originally estimated to cost Rs. 760 crores, the entire project would ultimately cost Rs. 2,190 crores (with Andhra Pradesh paying 1,108 crores).[23]

While the project ultimately did come to fruition, the celebrations of the first release of water for Chennai's water supply vividly capture the complex dynamics between state actors, political leaders, and the public narratives surrounding the success of such infrastructure projects. Media reports documented the long and costly journey to the project but celebrated the fulfilled promise of much-needed drinking water to meet Chennai's growing requirements. Politicians were ready for the customary public rituals to show that they had made good on their promise, ensuring that they would gain their share of the return of infrastructural political capital. Meanwhile, behind the scenes, local state actors and technical employers would have to do the labor of this public stagecraft. Engineers would have to work without pause to get the water to flow to the Andhra Pradesh–Tamil Nadu border on the day of the celebration, and as A. Mohanakrishnan would record,

> The water that had slowly trickled down to the border was temporarily held back for a short period of time until the Hon'ble Chief Minister Andhra Pradesh operated switch to open the shutter when the waters flowed through the measuring gauge in the deep reach at the border, with the Hon'ble Chief Minister, Tamil Nadu unveiling the Commendation tablet for the occasion. The entire crowd gathered, cheered and lined up along the bank to see the water flow through the Krishna Water Supply canal to infall into the Poondi reservoir [one of the main sources of drinking water supply for Chennai] 25 km off. (2011b, 75)

This carefully designed ritual captures a key dimension of the reproduction of state power. As with the successful implementation of the interstate agreement, the formal visible political rituals are the product of the intensive but rarely visible efforts of state employees.

The idealized narrative of the Telugu Ganga/Krishna Water Supply Project as a successful model of interstate cooperation thus masks a more entangled process of negotiation whose successful implementation was not predetermined. The decades since the first flow of water in 1996 have shown often limited and uneven successes when measured against the goal of providing twelve TMC of water for Chennai.[24] Both technical and political factors have posed obstacles to the delivery of water. The highest volume of delivery was 7.016 in 2009–10 (Mohanakrishnan 2011b, 43), and interviews that I conducted with officials from Chennai's utility company (Metrowater, August

17, 2016) confirmed that the city was not gaining the water resources they had hoped for from the project. In the initial years, technical complications with the canal posed problems for delivery. In more recent years, pressures from farmers within Andhra Pradesh and Andhra Pradesh's construction of new infrastructure projects have impacted the supply of water to Chennai (Ramadevi and Nikku 2008, 386). Such obstacles to the delivery of water to Chennai have been further complicated by the creation of the new state of Telangana in response to an ongoing popular movement in 2014. While the Srisailam Reservoir used for the Telugu Ganga Project remains within the newly bifurcated state of Andhra Pradesh, Telangana receives a portion of water from the reservoir. Since Telangana is not an official signatory to either the 1977 or the 1983 agreement, this has resulted in Andhra Pradesh arguing that a share of the water for Chennai should come from Telangana.[25]

Despite the serious disjunctures between the original promise of the supply of twelve TMC of water for Chennai and the limited delivery of water, there are important ways in which this interstate agreement remains a case of relative success. Most significantly, the bureaucratic and technical work of producing various mechanisms of communication and institutional cooperation has meant that continued disputes over the agreement are managed through negotiations rather than time-consuming and polarized forms of adjudication that occur when bilateral state conflicts over water-related matters become intractable. For example, in the context of the severe drought that placed Chennai's water supply in crisis in 2016, Tamil Nadu's chief minister made an unprecedented personal trip to Andhra Pradesh and was able to gain a public commitment from Andhra's chief minister that water would be released. While the release of water was far short of both the formal agreement and Chennai's supply needs, the significance of such attempts at reconciliation between the two states' interests should not be underestimated. Droughts, produced by complex configurations of human models of unsustainable development (and that may be potentially exacerbated by the effects of climate change) on one hand and the uncontrollable contingencies of nature on the other cannot be contained by the territorial boundaries of states. In crisis circumstances, when Tamil Nadu has faced extreme water shortages, it has been left with the task of trying to press its neighbors to enforce agreements with states that are managing their own water problems and crises. In such constrained circumstances, the politics of negotiations rests on intensified debates and conflicts over reservoir levels and the availability of scarce

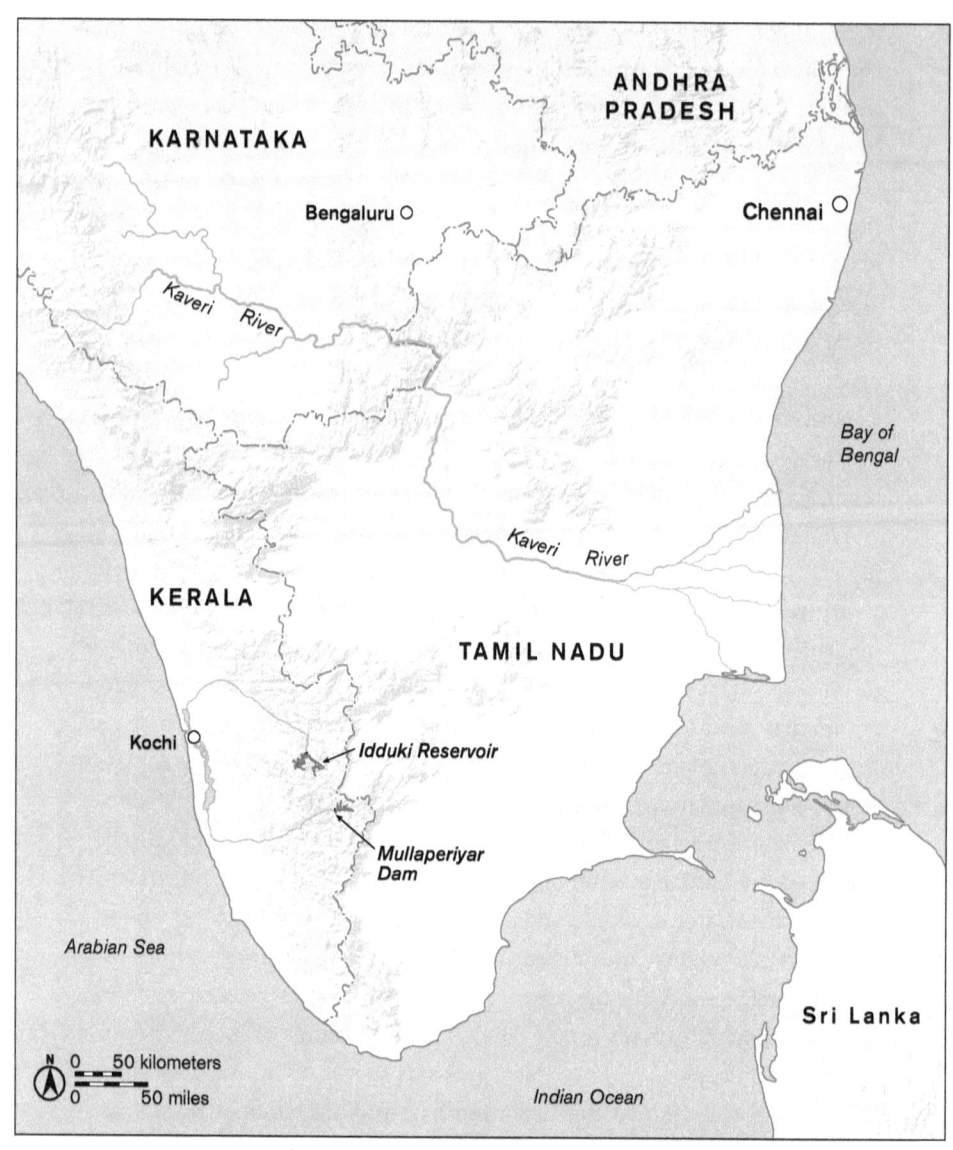

MAP 3.4. The Kerala Mullaperiyar Dam, presenting the location of the dam that has been a site of conflict between Kerala and Tamil Nadu. The dam is located in Kerala but managed by Tamil Nadu.

water resources that must be shared by the competing states in question. Technical experts employed by the state become critical actors in the management and mediation of such disputes as they provide the data that becomes the basis for negotiations. The case of the Telugu Ganga/Krishna Water Supply Project illustrates the ways in which the unglamorous hidden agency of technical experts and bureaucratic officials can provide a lasting institutional mechanism that can enhance interstate dialogue even when growing demands on water produce significant constraints on the implementation of formal agreements.

The ineffectiveness of the central state in providing either an effective national regulatory institutional framework for shared river governance or adequate machinery for the implementation of Supreme Court judgments or interstate tribunal awards has meant that state governments have resorted to the aggressive pursuit of their own interests. The result has been a continued hardening of the claims of state governments over water and water-related infrastructure in ways that deepen microprocesses of the centralization of state authority. Such dynamics are well illustrated in an interstate dispute between Tamil Nadu and its third neighbor, the state of Kerala.

Infrastructural Security and the Mullaperiyar Dam Conflict between Tamil Nadu and Kerala

On March 14, 2014, a team of engineers from Tamil Nadu's Public Works Department attempted to begin work on repairs of the floor of the Mullaperiyar Dam in Kerala. The engineers' work was halted after protests from Kerala's irrigation department despite the PWD's arguments that they had received permission from the Kerala Forests and Wildlife Department.[26] Kerala's objection was that the maintenance work was aimed at strengthening the dam—an issue related to a long-standing dispute between the two states that was under adjudication with the Supreme Court. The dispute was marked by a unique set of circumstances, in which the dam was located in Kerala but owned and operated by Tamil Nadu. Kerala had been raising issues regarding the safety of the dam and had been trying to decommission it, while Tamil Nadu was attempting to press for the dam to operate at a water-level height of 152 feet. On May 14, 2014, the Supreme Court delivered a verdict in favor of Tamil Nadu. However, the court decision has not produced either state or societal consent within Kerala. The result has been a legal resolution of this interstate dispute but a concurrent securitization of the infrastructure that has produced distrust and conflict between the two states.

The Kerala–Tamil Nadu dispute is shaped by both similarities and differences from the two other major cases of interstate water agreements and disputes that Tamil Nadu has had with its neighboring states. As with the Cauvery case, the contours of contemporary conflict have been shaped by the historical policies of the colonial state. However, in terms of the substance of the matter under consideration, the dam has stronger parallels with the Telugu Ganga Project, as it is an infrastructural project that must be managed between the two states. The issue at hand is not a dispute over water sharing but a dispute over the management of water infrastructure. Finally, underlying the overt focus on safety are political-economic interests in both states. In Tamil Nadu, an explicit reliance on water from the dam for both irrigation and drinking water in the context of systemic water insecurity heightened by Tamil Nadu's lack of control of river waters has hardened its attempts at gaining full control of the dam and its water height. Meanwhile, in Kerala, more nuanced interests in land and tourism in the context of liberalization intersect with its genuine concerns over the safety of the dam.

The Mullaperiyar Dam was constructed by the British colonial state in the second half of the nineteenth century and was specifically designed to divert waters from the Periyar River to serve the irrigation needs of Madurai in the Madras Presidency. After a prolonged set of negotiations, in 1886, the colonial state entered into a 999 lease agreement with the princely state of Travancore that allowed the British to lease the land needed, construct the dam, and maintain full ownership and control of its operation. This structure of ownership and operation was carried over into the postcolonial period, with Tamil Nadu owning and operating the dam under the terms of the 999 lease. As with the Cauvery case, the terms of this arrangement mirrored the relationship of power between the British state and the independent princely states in the colonial period. However, in contrast to the Cauvery case, there has been no political-economic conflict over the sharing of the Periyar River. Unlike Kerala, Tamil Nadu has a heavy dependence on water provided by the dam for irrigation needs as well as for hydroelectric power generation, which Tamil Nadu began in 1959 (Thateyus, Dhanaseeli, and Vanitha 2013). Tamil Nadu's dependence on resources from the dam has only intensified with the growing challenge of water scarcity. For instance, the city of Madurai, Tamil Nadu's third-largest city, has begun planning to use water from this source to meet its growing drinking water needs.[27] The agreement between the two states was then successfully renegotiated in 1970, with Tamil Nadu providing Kerala with fishing rights and Kerala agreeing to Tamil Nadu's right to construct facilities for power generation (Supreme Court judgment, 2014). As with the case of the Andhra Pradesh agreement, the two states were able to negotiate an agreement that merged their economic interests at the time.

The dispute between the two states was first sparked in 1979 when concerns about the safety of the dam began to take root in the public sphere in Kerala. Media reports in Kerala first began publicizing damage in the dam that was causing leakage (Madhusoodhanan and Sreeja 2010). The publicized damage, in conjunction with fears of the effects of an earthquake after a perilous dam failure caused by an earthquake in Gujarat in 1979, produced both societal and governmental concerns about the Mullaperiyar Dam. In response to a request from the government of Kerala, the Central Water Commission (CWC) conducted a series of inspections and subsequently instructed the Tamil Nadu government to engage in strengthening measures for the dam. At this time, the CWC recommended that the water-level height of the dam be kept at 136 feet until the strengthening work had been completed (Supreme Court judgment, 2014). This question of the height of the water level has become one of the central sources of contention in the dispute.

As with the Cauvery River dispute, years of adjudication, the politicization of the issue by both political parties and civil society organizations, and a complex set of political-economic factors has transformed this issue of dam maintenance into a decades-long dispute between the two states. The intensification of the conflict occurred in the late 1990s, after Tamil Nadu had completed the dam-strengthening measures and requested that the height of the dam be raised. The two states could not come to an agreement about raising the height of the water level, and Tamil Nadu eventually filed a petition in the High Court in 1998. This sparked a familiar chain of events comprising legal proceedings, state governmental maneuvers, increasingly inflamed political rhetoric, and public and social protests in both states until a final Supreme Court verdict deemed the dam safe and allowed Tamil Nadu to raise the water to 142 feet with the possibility of further raising the level to its earlier 2006 judgment of 152 feet. As with the Cauvery River dispute, existing scholarly work and media reports document in detail the history of the legal proceedings, the various political postures by political leaders, and protests by social groups. As I have noted, this interstate dispute is distinctive from the Cauvery case since it is centered on the management of infrastructure rather than river sharing. This distinction allowed the Supreme Court to claim full purview of the case in lieu of the tribunal process. Nevertheless, there are striking parallels between the two disputes in the underlying institutional and political-economic contradictions that play out in visible ways in court proceedings and political conflict in the public sphere.

The judicial and political conflicts between Tamil Nadu and Karnataka began unfolding in a context where policies of economic liberalization were accelerating with a significant shift at both the national and state levels in the 1990s. Such policies deepened the political and economic stakes in both states. As we have seen, in the case of Tamil Nadu, increasing pressures on water resources were created by continued urbanization. Both city needs for drinking water and industrial uses were intensified by urbanization and economic development. The intersection of such factors with the pressures of being a lower riparian state with unresolved water issues with its other two neighbors significantly increased the political stakes of control of the dam for Tamil Nadu. In the context of the three sets of interstate relations between Tamil Nadu and its neighbors, the Mullaperiyar Dam is the only case where Tamil Nadu has full ownership and control of the infrastructure in question. In the case of Kerala, while the public's safety concerns were genuine, the question of the water height also came with a set of less visible but important economic factors. The area around the Mullaperiyar Dam is a lucrative tourist area in an economy where a new embrace of globalization has heightened the importance of sectors of the economy such as tourism. A report commissioned by Kerala found that there would be a negative impact on revenues from tourism in the area (cited in Madhusoodan and Sreeja 2010, 21). The dam also has potential implications for the generation of hydropower for Kerala, as the Mullaperiyar Dam could potentially draw away water from a neighboring hydro dam in the district.[28] With the restraint on the height of the dam at 136 feet, numerous local businesses had cropped up in the 1980s and 1990s; raising the height of the dam meant a submergence of these businesses and a loss of the land. Policies of economic liberalization thus accentuated the stakes over two of the most scarce and valuable commodities—land and water.

The competing pressures that undergird the conflict over the dam once again stem from the ways in which state policies have promoted a set of economic policies at both the national and the local state level without providing an adequate institutional framework that can effectively manage economic and political relationships between states as they cope with the effects of such policies at the local level. Conflicts stemming from the scarcity of land and water are once again intensified by the weakness of regulatory mechanisms that can promote interstate cooperation. Critics of legal proceedings around such interstate disputes have expressed concerns about overreach of the Supreme Court in cases such as the Cauvery dispute and the Mullaperiyar Dam conflict (Iyer 2010). However, as with the Cauvery case, the court in effect stepped into a vacuum produced by the absence of effective national institutional mechanisms that could manage the effects of its economic policies. For instance, in the case of the Mullaperiyar issue, the evaluation of dam safety was conducted by Expert Committees set up by the central government only at the direction by the Supreme Court. Yet in 1979, the Government of India had set up the Dam Safety Organisation in the Central Water Commission "to locate causes of potential distress affecting safety of dams and allied structures and to advise/guide State governments in providing suitable remedial measure" (MWR 1987). As with the case of the River Boards Act in the Cauvery case, this regulatory central government institution did not have the substantive mechanisms that could allow it to provide a foundation for reconciliation between Kerala and Tamil Nadu. Once again, the formation of Expert Committees and technical discussions of safety could only occur after processes of adjudication and mobilization around such judgments had politicized the technical data.

The institutional deficiencies of the state's regulatory frameworks that have catalyzed a more active role of the Supreme Court at the national level have also accentuated the divide between legal and sociopolitical terrains of power and authority. In contrast to the Cauvery Tribunal award, the Supreme Court judgment has been both binding and enforceable because of Tamil Nadu's ownership and control of the dam. However, the formal resolution of the case has not produced a corresponding political or societal form of consent to the judgment in Kerala. Mobilization over the dam's safety has been shaped by a complex mix of political interests, social movements, and media narratives. The politicization of dam safety, for instance, must be understood in the context of a broader national sociopolitical field in India that witnessed growing environmental movements mobilizing against the impact of the dam.[29] The initial petition filed in Kerala to stop the raising of the dam height level was filed by a civil society environmental organization, the Mullaperiyar Environmental Protection Forum (SCI 2006). State governmental action in Kerala, including a 2006 ordinance that sought to nullify the 2006 Supreme Court verdict (by limiting the height to 136 feet after the 2006 decision allowed for the water level to be increased), is often in part a response to social and political pressure from movements and actors within civil society. Similarly, farmers' protests in Tamil Nadu have broken out at various points when farmers have been faced with the threat of water supplies from the dam being denied for their crops.[30]

Indeed, pressure from societal actors, including the media, can restrain state governments from opening up space for interstate cooperation. A vivid example of this was evident when Kerala's newly elected chief minister Pinarayi Vijayan made a public statement suggesting that the question of the safety of the Mullaperiyar Dam was settled by the Supreme Court and that Tamil Nadu and Kerala should resolve these issues through bilateral talks. The comments were immediately politicized within the public sphere with the opposition party's leader saying that the chief minister's "new stand was against the pulse of the people of the State."[31] The chief minister was soon compelled to reverse his position, and public contestation over dam safety has continued in both states. The political limits on the agency of state officials at the local level play a significant role in shaping broader outcomes at the interstate and national level. The representation of the "pulse of the people" foregrounds the embeddedness of states within civil society. Local state actors respond to varying political and social pressures. Farmers (who are themselves a varied group) require irrigation water for their livelihoods, while urban residents in cities need drinking water. Meanwhile, formal civil society institutions such as NGOs and the media (often located in cities that have become powerful sites of investment in postliberalization India) have an important impact on public discourses. State actors weigh these competing demands based on political and electoral calculations.

In the absence of productive interstate negotiations of the kind attempted by the Cauvery Family initiative, such political and societal pressures, along with real concerns of security and livelihood in both states, remain latent sources of conflict. In periods of crisis, the absence of adequate interstate institutional frameworks of cooperation can produce serious consequences.

The underlying volatility of the conflict was vividly brought to light during historic floods in Kerala in 2018, when the height of water in the dam became a point of contention in the midst of severe crisis. The 2018 floods in Kerala have been widely described as the worst floods in the state since the historic flood in 1924. The floods claimed at least 417 lives and produced widespread destruction of livelihoods and property in both rural and urban areas. While the flooding of Kochi's (formally known as the city of Cochin) major airport provided a visual symbol of urban flooding (with the airport closed for a period of two weeks), the entire state was severely affected. Idduki District in the western Ghats was particularly hard hit. A combination of flooding, landslides, and communications and power failures fully isolated the district and compounded the devastating effects on the area. In the midst of this calamitous set of events, the Mullaperiyar Dam, which is located in Idduki District, became a source of heightened anxiety in Kerala. While the Mullaperiyar Dam can draw 2,300 cusecs (cubic feet per second) from the reservoir for irrigation, the inflow of water had reached 20,508 cusecs. The matter at hand was of critical import for Idduki, as the overflow of water from the dam would drain into the Idduki Reservoir.

The institutional and political dynamics of the crisis followed the pattern of earlier processes. During the height of the crisis, consistent with the institutional pattern of interstate governance, the Supreme Court intervened and asked Tamil Nadu to reduce the water level from 142 to 139 feet.[32] Tamil Nadu's chief minister responded with the defensive reaction that the dam was safe at 142 feet and was not the cause of the floods, pointing to the fact that the dam's water level had reached only 140 feet. Kerala, on the other hand, blamed the lack of centralized coordination and the dam's water levels for exacerbating the floods with sudden releases from the dam that forced the release of water from the Idduki Reservoir.[33] The inadequacy of the interstate institutional framework of cooperation was brought into sharp view at the height of the crisis, as the Supreme Court directed the National Committee for Crisis Management to review the possibility of reducing the dam's water level. Directing the various governmental levels to respond effectively, the Supreme Court acknowledged its own inadequacy in the midst of the crisis. As one report noted, "The court said it need not be overemphasized to state it was not an expert to issue any kind of guidelines to manage a situation of the present nature."[34] While Tamil Nadu later accepted the National Committee's recommendation to lower the water level to 139 feet, the significance of the political tensions and institutional regulatory shortfalls that I have been analyzing come into stark view through the serious consequences in the context of this kind of disaster management.

Such conflicts over water infrastructure, as I have noted earlier, are not inevitable—nor do they have to lead to insurmountable animosity between neighboring states. Rather, they are products of an accumulated history of state policies and institutional deficiencies. Consider the complex nature of the relationship between Tamil Nadu and Kerala. Shared economic ties between Tamil Nadu and Kerala, for instance, have in the past provided the underpinning for Tamil Nadu's effective use of a boycott during the dispute of the Mullaperiyar Dam in 2006 (Madhusoodhanan and Sreeja 2010, 20). Kerala's reliance on low-cost agricultural products from Tamil Nadu meant that the boycott had a serious impact on its population. While this serves as an instance of the deterioration of relations between the states, it also highlights the various relationships and mutual forms of dependencies that exist between them. The two states have cooperated over other water-sharing issues, as they relied on each other for water sharing even during the prolonged dispute over the dam. They have also constructively worked together in sharing water through the Parambikulum-Aliyar Project (PAP). As with the case of the Telugu Ganga Project, the PAP has been managed by an institutional structure, the Joint River Water Regulation Board, which meets regularly in ways that keep open lines of communication between the states.[35] Meanwhile, the Siruvani Dam in Kerala has served as a source of drinking water for the major city of Coimbatore, in Tamil Nadu. The two states have been able to engage in negotiations that build on such mutual dependencies. In the context of drought periods, the states have struck deals to release water from PAP to Kerala and from the Siruvani Dam for Coimbatore's drinking water supply needs.[36] Such spaces for interstate cooperation are increasingly critical as the conjunctural effects of economic development, climate change, and natural stresses will continue to provide acute stresses on shared resources between states.

When interstate cooperation and dialogue stall or break down, as they have in the context of the Mullaperiyar Dam, the absence of effective national institutions needed to mediate relationships produce a more volatile form of securitization of water infrastructure. State governments turn to the coercive dimensions of state power, such as the police and security forces, when national institutional mechanisms for conciliation are absent. The centralization of local state governmental authority in this context begins to take on characteristics of the security state. For instance, the conflict between the two states over the dam has not been confined to courtrooms or ministerial meetings but has played out through increasingly tense relationships between local bureaucratic and technical officials who are tasked with the management of infrastructure. While the dam is fully owned and operated by Tamil Nadu, the Kerala police have guarded the structure. During periods of tension in the course of the conflict, this has meant that employees of Tamil Nadu's Public Works Department have encountered hostility— in some cases through attacks on their technical work and in rarer instances through attacks on their physical well-being.[37] As the collection of data on the safety of the dam has become politicized, officials from both state governments have sought to limit access of technical officials and research teams from the other side.[38] The intensity of mistrust at the local level has been such that individuals have also been accused of serving as spies.[39]

Such animosity at the local level poses a significant challenge to the continued interstate cooperation required between the states after the legal resolution of the water-level height in the Supreme Court. After the Supreme Court verdict, Tamil Nadu filed a request for the deployment of security from the Central Industrial Security Force to facilitate the operation of the dam, and Kerala, in response, made plans to open a full-fledged police station near the dam.[40] While Tamil Nadu later withdrew the request after it was rejected by the Supreme Court, the action illustrates the securitization of the dam's operation.[41] As with the Cauvery case, the terrain of law and adjudication does not provide a sufficient basis for reconciliation when it is not accompanied by concrete mechanisms for future cooperation. Kerala has continued to make public protestations about the water-level height (which, at the time of writing, is at 142 feet). Meanwhile, Tamil Nadu continues to press for raising the water-level height to its earlier full height of 152 feet, given its escalating challenges of water scarcity during drought periods.

This case of water infrastructure has become the basis for competing conceptions of security. In Kerala, entrenched fears about the safety of the dam have become woven into public views of the security of the state's residents. In Tamil Nadu, continual public anxieties about water supplies have become fundamental elements of the security of both urban and rural residents who rely on water from the dam for drinking supplies and agricultural needs.

The securitization of the dam through policing practices by representatives of both local state officials and local communities across the border fills in the vacuum of sustainable and effective institutional practices that can build trust and cooperation between the two states. The establishment of such institutional mechanisms is crucial to prevent infrastructural politics from becoming overdetermined flashpoints that overwhelm the mutual dependencies and shared interests that undergird interstate relations in postliberalization India. In the face of weak national regulatory mechanisms, local state governments increase their claims of sovereign authority over water and water-related infrastructure.

Tamil Nadu's riparian position has placed it in a geographical context where the management of water resources has enmeshed the state in ongoing interstate negotiations with all three of its neighbors. Taken together, the three major cases of interstate negotiations and disputes examined here reveal the dynamics of federalized state authority over water in the postliberalization period. The historical legacies of both colonial rule and the impact of planned development in the early decades of independence have produced enduring political-economic structures that place political strains on the federal management of water resources in southern India. These strains have been deepened in the postliberalization period—water-intensive irrigation needs now compete with processes of urbanization and national policies that have encouraged states to accelerate power production through sources such as hydropower and aggressively pursue investment in new industries that add new demands on water resources. Such economic policies have intensified interstate competition for water resources without institutionalizing regulatory mechanisms that can promote interstate cooperation over shared resources.

The dynamics of centralization and decentralization play out in complex and contradictory ways. The hollowed-out nature of the regulatory state at the national level has increased the importance of local state actors—a process that is a byproduct of state incapacity at the center of the federal system and not a sign of a designed set of policy reforms or a democratic process of decentralization. Meanwhile, the dynamics of interstate conflicts and electoral and political pressures on state governments have meant that state governments have intensified their claims of authority on water and water-related infrastructure. Such claims are subtle signs of new forms of centralized authority over water resources that are taking root at the local level.

The cases analyzed here point to a nuanced set of processes that are shaping the centralized control of water at the local level. We have seen that interstate disputes conceal more complex conflicts between different users within states. The territorial nature of interstate disputes and negotiations in effect conceals the crucial differences of interests between different water users in liberalizing India. The pressures on water resources and the phenomenon of water scarcity are themselves shaped by changing patterns of urbanization, investment, and development that are effects of policies of economic reforms. In the context of reforms that have accentuated a city-based model of growth, the realm of local urban governance becomes a central site for an understanding of the governance of water in the postliberalization period.

CHAPTER 4

Regulatory Extraction, Inequality, and the Water Bureaucracy in Chennai

CITIES IN INDIA REPRESENT CRITICAL SITES FOR AN UNDERSTANDING of how institutional reforms have shaped the governance of water in the postliberalization period. Policies of economic reform have intensified the political and economic power of metropolitan urban centers. Reforms of the governance of water produce a redistribution of state power that is shaped by this ascendancy of a city-based model of development. On the one hand, reforms expand the centralized authority of some city-oriented agencies of the water bureaucracy. On the other hand, policies of decentralization target small towns and rural areas in ways that both reflect the political and economic weaknesses of these localities and intensify the control of local state governmental authorities over these areas. The realm of urban governance thus tells us a story about the postliberalization state—one that speaks to a broader set of changes in the underlying relationship between the city, small towns, and rural areas.

While urbanization has been accelerating and small towns in India have been growing in both number and economic importance, the major metropolitan cities and their environs—Mumbai, Delhi, Bengaluru, Kolkata, Chennai, Hyderabad, and Ahmedabad—remain the central sites for the implementation of economic policies of liberalization, the concentration of wealth and investment, the centers for population growth, and the locus of political power. In this context, metropolitan cities do not represent bounded urban sites that are limited to the territorialized administrative boundaries of metropolitan cities. Cities are microcosms of global-national patterns of reform and are spatial sites that are deeply imbricated in interconnected social and economic relationships with both the urbanizing communities that populate their immediate peripheries and distant rural localities that appear far from their borders. These shifts take place in a context where increasing demands on often scarce water resources for drinking water, agricultural, and industrial needs in the postliberalization period are deepening the pressures on water bureaucracies. Urbanization has been producing new strains on scarce water resources and water-related infrastructure.

Consider the following example of some of the challenges that Chennai has faced in the context of oscillating pressures of floods and droughts. Chembarambakkam Reservoir is one of the three major reservoirs that supply water to the city of Chennai. In December 2015, delays in opening the sluice gates of the reservoir were widely reported to have been a key factor in producing historic flooding in Chennai during a four-day period of unprecedented heavy rainfall.[1] Two years later, in a period of unprecedented drought, the assistant engineer who was responsible for managing the gates pointed to the depleted reservoir and reflected on the stress and anxiety he had experienced during the flood. He had spent ten days monitoring the water levels on his own while facing the grave possibility that a breach in the reservoir would cause a catastrophic flood. He recounted the consistent phone calls from governmental officials and the fear he felt that he would be blamed if the reservoir were breached (interview and field visit, January 19, 2017). The reservoir lies at the edge of Chennai and is surrounded by numerous small towns and urbanizing localities that are classified as the "peri-urban" areas that often appear as the unplanned outgrowths of metropolitan cities in India. Had the reservoir collapsed, the flooding would have been catastrophic for these localities. In the context of the dwindling water supplies of the drought-affected reservoir that we were looking at, the engineer's memory of the flood was laced with irony.

This anecdote encapsulates the entangled story of state practices and the patterns and contradictions that shape and constrain governance over water and water infrastructure. The misjudgment of local state officials on the opening of the sluice gates that acerbated the flooding in Chennai points to the serious implications of bureaucratic action—and inaction. The dwindling levels of water in the reservoir, two years later, point to the strains on the state during periods of water scarcity, as it must manage the growing demands for water from a heavily urbanized city whose boundaries have been steadily expanding. Underlying this account of the pressures during times of what appear to be "natural" calamities of flooding and drought are deeper structural pressures that various models of urban development have placed on the city. Expanding development on wetlands has increased the severity of the floods, as there are no natural drainage areas to catch the water. These processes of urban development have in turn increased demands for water for drinking supplies and agricultural needs, as well as for industries, as private investment has expanded in Tamil Nadu over past decades. Meanwhile, broader human developmental activities that may be shifting weather patterns and producing natural phenomena such as failed monsoons create unpredictable strains on the state and on its ability to manage competing demands in times of water scarcity.

As I stood with a group of engineers from the Public Works Department listening to them discuss the weight of these strains, another assistant engineer commented on how much the surrounding areas had changed over the past two decades. Pointing to these areas, the engineer commented, "This was all agricultural before. In the 1990s, the government said, 'Let it become urbanized'" (interview and field visit, January 19, 2017). The assistant engineer was suggesting that the government began tacitly withdrawing support for the surrounding agricultural communities and in effect allowed the urbanization to occur. What appears as a disorganized urban outgrowth of the city of Chennai was in fact shaped by state decisions on the allocation and withdrawal of resources. The offhand comment, "The government said, 'Let it become urbanized,'" provides a microinstance of the centralized authority of state governments over the reallocation of water resources. We see here the often hidden intentionality of the withdrawal of the state—in this case through stopping the procurement of agricultural products that often sustains agricultural communities.

In this example, the removal of state support did not embody a transition from the centralized state support of the developmental state to the kind of decentralized model of governance that is conventionally associated with reforms. Rather, the withdrawal of support reflected a reorientation of centralized state authority and a shift of the state's resources away from the agricultural communities. The state does not abandon but restructures its welfarist framework—it produces a redistributive shift that reallocates water resources to privileged groups within wealthier urban centers. This accentuates long-standing socioeconomic inequalities within urban centers while deepening new divides between larger urban areas and smaller rural and urban towns. Institutional reforms in the process produce or intensify inequalities, such as those of class, locality, caste, and gender.

Institutional reforms provide the mechanisms of regulatory extraction that produce differential access to water resources and intensify these relationships of power both within and between urban and rural communities. Institutions are the heart of governance, and they have the capacity to ameliorate, reproduce, or intensify inequalities. In the postliberalization era, patterns of inequality are produced and intensified by institutional reforms that give some city-oriented state agencies new forms of authority while weakening other bureaucratic agencies. Policies of reform in this context produce an institutional redistribution of authority rather than a framework of decentralized or participatory governance. Meanwhile, policies of decentralization tend to target small towns and rural areas that are politically and economically weaker than metropolitan cities. Reforms in effect produce a form of differential decentralization that embodies these underlying relationships of power. In this process, regulatory reform is transformed into a process of regulatory extraction that encodes relationships of power both within and between urban and rural communities.

Reforming Chennai's Water Bureaucracy

Chennai's water bureaucracy has experienced significant shifts in its institutional landscape in the postindependence period. The state government embarked on a significant program of reform, the Tamil Nadu Water Resources Consolidation Project, through a $282.9 million loan funded by the World Bank from 1995 to 2004 (WB 1995b; the total project cost was $491 million). The project was one of only three set up in states that sought to implement the Bank's new integrated water resources management approach in the early stages of liberalization in India (WB 1995b; the other states were Odisha and Haryana). The project engaged in a comprehensive reorganization of the management of Tamil Nadu's water resources along the lines of conventional dominant global models that have emphasized the modernization of irrigation systems, technocratic improvements in the management of water resources, and the creation of participatory frameworks through Water Users' Associations. A key dimension of the reorganization was a shift to water planning based on river basins that would cut across various water users. Indeed, as we will see, the project has reshaped institutions as well as created new organizations and state practices at both the state and local community level. More significantly, such shifts toward new regulatory mechanisms have in turn produced new forms of centralization that complicate the policies and rhetoric of decentralization. First, reforms have mirrored broader patterns that have characterized this kind of institutional regulatory transplant. These new regulatory practices have simply been molded onto existing institutional relationships and practices through a form of regulatory "shell" that is often the reality of global institutional transfers (Dubash and Morgan 2013). In this context, new regulatory practices have been shaped by the relationships of power between the various institutions that make up the water bureaucracy and by the underlying political-economic structures of development. These are the key dynamics of reform rather than the idealized global norms of accountability and technocratic and participatory efficiency of global institutions such as the World Bank. Second, and more significantly, policies of institutional reform have provided the means for new forms of centralization. The Bank's turn toward an emphasis on state accountability and ownership of reforms has produced the institutional scaffolding—through policy, legislative, and organizational changes—that has consolidated modes of centralized state authority over water.

Within Tamil Nadu's institutional landscape, the Public Works Department has retained control over irrigation as well as over the regulation and storage of water. Tamil Nadu's Public Works Department, in keeping with the historical weight of its institutional authority, is the only such department in the country with control over irrigation. This preservation of authority has meant that the PWD has remained a leading institutional actor within the water bureaucracy. However, the PWD's institutional monopoly has also been weakened by various phases of institutional reform. The first phase of institutional restructuring that significantly shaped Tamil Nadu's water bureaucracy took place in the 1970s. A major set of institutional changes in the 1970s restructured the Public Works Department (PWD) through the splitting off of drinking water supply needs for both rural and urban areas. Chennai's water supply and sewerage needs were placed under the newly formed Chennai Metropolitan Water Supply and Sewerage Board (CMWSSB, more commonly known as Metrowater) in 1978, and the state's water supply was placed under the purview of the Tamil Nadu Water Supply and Drainage Board (TWAD) in 1970. Meanwhile, the Chennai Municipal Council has also remained an important actor in this field, as it has maintained control over storm drainage management as well as the management of urban development, which has a direct impact on water management. Water management in the state, once under the sole purview of the historical, imperious institution the Public Works Department, is now shaped by a mosaic of institutions with distinctive yet interconnected and overlapping functions (see figure 4.1).

Institutional reforms that have taken place since the 1990s have continued their focus on regulatory reform. The emphasis on institutional restructuring has in large part stemmed from the fact that water resources have long been overutilized in the state. Consider, for instance, the assessment of the external consultancy firm that was hired for Tamil Nadu's Water Resources Consolidation Project. The firm recommended that the project focus on "upgrading technical and management skills" since "Tamil Nadu has developed its surface and groundwater resources almost to physical limits" (WRO 1996). The PWD was consequently reorganized, and the Buildings and Water Resources wings of the department were split into separate organizations. Given that water resources in the state were already overexploited, World Bank–sponsored reforms focused on institutional reorganization that could enhance the management of water resources. As the stated objectives of the Water Resources Consolidation Project noted, "Under the project, a formerly construction oriented Public Works Department (PWD) would be refocused and strengthened as a state water agency responsible for multiuse water planning and for providing irrigation, drainage, flood control and bulk water supply services. Expenditures would be refocused to emphasize maintenance and modernization of existing facilities, and beneficiary participation linked with cost recovery would be integral to the service improvements" (WB 1995a, 1). To that end, the project successfully led to the establishment of a new regulatory organization within the PWD, the Water Resources Organisation. The reorganization along four regions in the state (Chennai, Madurai, Trichi, and Coimbatore), each with its own chief engineer, was specifically aimed at deepening the decentralization of water governance.[2] The focus of the Water Resources Organisation was specifically geared toward what the government would term "the effective management

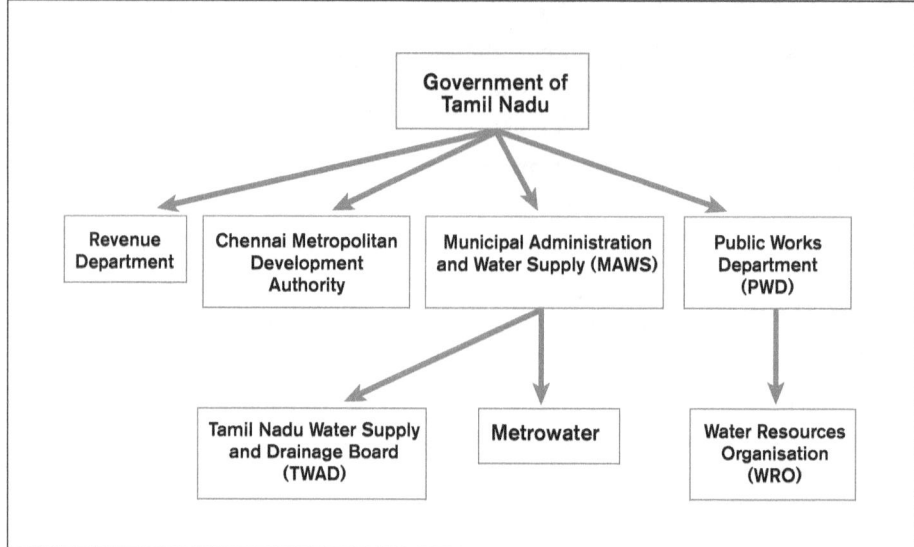

FIGURE 4.1. Chennai's Water Bureaucracy, identifying the major bureaucratic organizations and their administrative reporting lines in Chennai's water bureaucracy

and distribution of Surface and Ground Water for its optimum utilization in a rational and scientific manner by all water using sectors."[3]

Despite these endeavors aimed at improving and rationalizing institutional practices through regulatory reform, water governance has been shaped by the historical legacies of bureaucratic organizations as well as the domestic political and economic priorities of Chennai's regime of governance. Consider, for instance, the ways in which periodic flooding in Chennai has been affected by relationships between key bureaucratic organizations. The rapid pace of urbanization in Chennai began in the 1970s (see table 4.1).[4] In a precursor to the historic 2015 floods, Chennai experienced extreme flooding in 1976 in large part due to drainage problems as a result of urbanization (MMDA 1993, 2–3). Chennai's stormwater drainage system is laid alongside the edge of roads rather than underground, and there are "numerous cross-connections between the foul and stormwater systems" (6-2). While the newly formed Metrowater was given authority over Chennai's sewer system, stormwater drains remained within the purview of the Madras Metropolitan Corporation. As a Madras Metropolitan Development Authority

TABLE 4.1. Urbanization in Tamil Nadu

Year	Urban population (millions)	Share of urban population (%)	Decadal urban growth rates	Rural population added during the decade (%)	Urban population added during the decade (%)	# of urban towns
1901	2.72	14.15	—	—	—	133
1911	3.15	15.07	15.51	—	—	162
1921	3.25	15.02	8.86	61.57	13.63	189
1931	4.23	18.02	23.40	56.48	53.28	222
1941	5.17	19.7	22.30	66.26	33.74	257
1951	7.33	24.35	8.39	43.92	56.08	297
1961	8.99	26.69	22.59	53.56	46.44	339
1971	12.46	30.26	38.64	53.75	46.25	439
1981	15.95	32.95	27.98	51.63	48.37	434
1991	19.07	34.65	19.59	58.05	41.95	469
2001	27.48	44.04	44.06	-28.41	128.41	832
2011	34.95	48.45	27.16	23.29	76.80	1097

SOURCE: Government of Tamil Nadu, "Urban Scenario in Tamil Nadu Census of India," www.tn.gov.in/cma/Urban-Report.pdf and Census of India 2011 (Chennai 2021 population was estimated at 8.65 million).

report on stormwater drainage noted, a proposal to transfer the management of stormwater drainage to Metrowater was opposed by the Madras Metropolitan Corporation (MMC) because the MMC was in charge of road infrastructure, which was in turn dependent on the functioning of the roadside drainage system, given that the system is laid along the edge of the roads rather than underground (6-2). As the report went on to note, stormwater drainage was underfunded and was the lowest priority for the MMC (6-3). The report notes similar smaller institutional fractures in the management of water-related matters between the MMC, Metrowater, and the PWD. For instance, all three entities were embroiled in small-scale disputes over the control and management of arterial drains. While arterial drains that would flow into rivers were the responsibility of the PWD, the MMC and the PWD were competing for authority over the management of drains that were receiving outfall from stormwater drains. Meanwhile, according to the report, foul sewage, which was under the purview of Metrowater, was at times "discharged illegally by users to arterial drains and to channels maintained by

PWD" (6-3). Such forms of competitiveness and the impetus that bureaucratic organizations feel to protect their spheres of authority are, of course, an intrinsic element of all institutional landscapes. However, they also shed light on the ways in which the historically contingent material nature of infrastructure can both deepen and be shaped by such institutional cleavages. In this instance, the specific kinds of connections between the stormwater drainage system, the sewer system, and roadways deepened the obstacles to reorganizing the management of the stormwater drainage system in ways that could provide more effective flood control.

Meanwhile, the deeper underlying institutional division that is noted in the government report has to do with the primary cause of heightened problems with the stormwater drainage system, which it identifies as the "rapid pace of urbanization" (MMDA 1993, 2-3). While the report reproduces a familiar state discourse on the problem of "encroachments" on rivers that affect flood drainage, it also points to problems with the construction of the Mass Rapid Transit system, which had just begun in the early 1990s, as well as the spread of "impermeable surfaces," such as buildings, roads, and pavements, which were intensifying the threat of flooding and which, over two decades later, would lead to the historic 2015 flood that brought the entire city to a standstill.[5]

What is of critical significance in this story of institutional cleavages over infrastructure management is the ways in which the emphasis on decentralization in effect provides both the institutional and political space for the policies of urban development that strain the city's water resources and infrastructure. While decentralization in this case was targeted at the water-related entities of the PWD, this process was accompanied by a centralization of power within other components of the state government, whose developmental agendas were being shaped by policies related to liberalization. The fraught institutional cleavage in this instance lies with the separation between urban developmental decisions and activities placed under one of the major governmental bodies in the city, the Chennai Metropolitan Development Authority (CMDA) on the one hand and the various organizations that make up the water bureaucracy on the other. Institutional reforms, such as the Water Resources Consolidation Project, that have been supported by global models of water management have treated the water bureaucracy as a closed system that can be isolated from state structures and policies that regulate land and development in the city.

Consider, for instance, the internal policy and strategy discussions of the Water Resources Organisation that were instituted through the Water Resources Consolidation Project reforms. The department's evaluation of its policies and strategies focused on a broad and nuanced understanding of the structural problems posed by urban developmental practices. The most significant concerns of the WRO's report were focused on the pressures of urbanization being produced by business interests in the real estate market. As the report noted, the encroachment of water bodies around the city, which were intensifying the strain on the city's management of water resources, were being affected by fact that "the lucrative prices offered by the real estate businessman for the urban lands lure the agricultural land holders to sell their agricultural lands for housing purposes" (PWD 1994, 172). While the Tamil Nadu government had instituted laws to regulate the conversion of agricultural land to residential housing, the report noted that "in spite of these steps taken by the Government, the conversion of wetlands goes on in view of the high prices offered for the land" (173).[6] The division in institutional interests between the planning authorities of the government and the water bureaucracy are well illustrated in this acute assessment of the transformation of the real estate market that was taking root as India began liberalizing its economy. Chennai's IT corridor, for instance, was built across wetlands, while expanding residential developments have substantially encroached on floodplains in the metropolitan area. This process has continued to expand as urbanization has extended beyond the borders of the city. As one news report noted, "Planning permissions inside the Chennai Metropolitan Area (CMA) are based on whether the builder gets 'No Objection' certificates from Metrowater, electricity boards, traffic and fire services. However, a promoter building a 27-storeyed complex beyond Uthandi, outside CMA limits, will approach the directorate of town and country planning and local authorities who don't thoroughly scrutinise the applications, says the official."[7] Indeed, in 2021, Tamil Nadu is now one of the most urbanized states in the country, with 48.4 percent of the population living in urban areas.[8]

The WRO's report further assesses the strains that unplanned urbanization have placed on the city's water sources and supply in light of the reform policies that have actively encouraged state governments to attract and compete for investment. Growing industrial investment in the city began to intensify stresses on the city's water supply, particularly with competing demands from industrial and residential consumers. In response, the Water Resources

Organisation recommended both prioritizing drinking water over industrial needs and creating new regulations on the establishment of new industries based on their water consumption needs that would "permit only industry that doesn't require large quantity of water" (PWD 1994, 163).

This synopsis of WRO policy evaluations and recommendations reveals a more heterogeneous bureaucratic field than do conventional unitary portrayals of India's bureaucracy either as essentially corrupt and in need of reform or replacement by private sector management or as a simple bureaucratic arm of private sector interests. The WRO, in this context, was attempting to execute its regulatory function. However, such regulatory attempts were foreclosed by the state government's centralized push for investment in accordance with the broader global-national norms of liberalization.

Such processes point to the deeper internal structural contradictions of the global norms of economic and institutional reforms that are transplanted to contexts in non-Western countries. In this case, for instance, the Water Resources Organisation's recommendations reflect a bureaucratic organization that is working effectively and that is trying to manage the strains of developmental demands on scarce water resources. The WRO, in effect, is attempting here to perform its regulatory functions. However, the central obstacles to the organization's effective institutional practice in this case lie not in any intrinsic bureaucratic dysfunction within the organization but in more powerful sections of the state bureaucracy that are pushing forward with urban developmental practices that have become highly lucrative in the postliberalization period.

Consider another example of the internal contradictions within state bureaucracies. In response to the effects of rapid urbanization on water bodies, the Government of Tamil Nadu passed an order to regulate and restrict the conversion of agricultural land to housing sites (PWD 1994, 172; the law was passed in 1991).However, as the PWD's Water Resources Organisation would note in an internal report, while planning authorities needed prior agreement from the Agricultural department for such construction and were specifically meant to avoid building on wetlands, "In spite of these steps taken by the Government, the conversion of wetlands goes on in the view of the high prices offered for the land" (173). Equally significant was the fact that the order specifically exempted the construction of government buildings from this regulatory process (IWS 1994). The result has been that a number of "encroachments" on water bodies in and around the cities have been due to the construction of government buildings that cannot be removed (interview with director, Centre for Water Resources, Anna University, January 11, 2017). The water bureaucracy is placed in an institutional environment in which they have little control over the macroeconomic and developmental decisions that have been systematically straining the water supply of the city. There is a fracturing of the regulatory state that produces this structural contradiction. State water resource management organizations are tasked with regulating the city's water supply in a broader regulatory regime that does not enforce formal regulations of land use and urban development. A recognition of these contradictions is markedly absent from global and national policies and discourses of bureaucratic reforms that have been a central part of liberalization in India.

In practice, this has meant that the bureaucratic organizations concerned with flooding have targeted encroachments by groups that are socioeconomically marginalized and politically less powerful than state governmental organizations invested with power over land and development. Anthropologists Karen Coelho and N. Raman (2013), for instance, have argued that the government's water body restoration projects in the city have focused on the eviction of poorer, vulnerable communities through slum clearance activities while continuing with accelerated large-scale developmental activities, which are the primary cause of environmental degradation in the city. They note that the Tank Encroachment Act (2007) "ignored or reversed long-established policies guiding slum clearance in the state of Tamil Nadu and vested unprecedented powers in the Public Works Department (PWD) and the District Collector's Office to effect evictions, entirely bypassing the Slum Clearance Board. The thrust to revive storage capacity in water bodies received powerful political backing by the state's ministers, legislators, and members of parliament in the late 2000s" (2013, 151). In this context, the PWD has acted as an enforcement arm for the state's developmental agendas in ways that have transformed metropolitan city environmental agendas into the kind of class-based endeavor that has been framed through the exclusion of socioeconomically subordinate groups across the country (Fernandes 2004). Aspects of internal departmental strategies for the management of water resources do also tangentially echo class-based concerns surrounding the impact of the "encroachments" of the urban poor with particular concerns about "slum dwellers and the floating population from the other parts of the state, polluting the environment" (PWD 1994, 77). What are in fact marginal references in the internal planning discussions of the WRO nevertheless are transformed into the politically viable default target of the state. It is here that we see how the distribution of institutional power matters. State administration structures governing land and development outweigh the regulatory potential of the water bureaucracy.

The examples of the divergence between the reforms aimed at decentralization in the water bureaucracy and the intensification of state governmental authority through formal and extralegal modes of urban development are not isolated instances. Nor are they simply evidence of the corrosion of governance by domestic politics. Rather, they are symptoms of processes of reform that consolidate state authority through the growing political and economic power of metropolitan cities. This centralization of power through the space of the city is an inherent part of the twin processes of economic reform and institutional decentralization; the regulatory state is transformed into a mechanism of regulatory extraction that is encoded in processes of institutional and economic reform.

Inequality, Regulatory Extraction, and the Redistribution of Bureaucratic Authority

"There is no PPP [public-private partnership] model here. We don't want private financing. Water is a public good" (interview, August 17, 2016). This emphatic assertion by a senior engineer at Metrowater represents a sharp deviation from dominant understandings of the impact of reforms on the water sector. Debates on water reforms in India have often been shaped by a preoccupation with the effects of privatization. Critics of liberalization have called attention to the dangers of privatizing water resources, and proponents of reforms have largely focused on the need to harness private sector participation in the development of water-related infrastructure. Indeed, Tamil Nadu has often been held up as an example of a state that has taken the lead in the privatization of the water sector. Processes such as the implementation of reforms within utilities such as Metrowater, the establishment of the Tamil Nadu Urban Development Fund, and the increasing reliance of Chennai on the private supply of water have led to an understandable emphasis on the ways in which privatization has transformed the management of water resources in Chennai (Coelho 2005a, 2010; Gopakumar 2012 Mahalingam, Devkar, and Kalidindi 2011). However, a sole focus on the logic of privatization also masks more complex sets of relationships between the state, civil society, and private capital—a relationship that consolidates new forms of centralized state authority.

The adamant rejection of a model of private financing by the Metrowater engineer represents more than an idealized assertion of the publicness of the water utility's function or a political defensiveness against the role of the private sector. It captures the ways in which various facets of state institutions and state power shape the creation of water markets even within a state that represents a strong case of the implementation of reforms and policies of privatization.[9] State practices actively shape the formation of water markets through the regulation of resources. An analysis of Chennai's water supply provides an in-depth understanding of this remaking of state power and markets in the context of global processes of reform that are shaping India's society and economy. Such an analysis moves beyond a city-centered story of urban inequality and requires a deeper engagement with the ways in which state practices emerge from, intensify, and manage complex inequalities both between and within urban and rural localities. Chennai's water supply is the product of historically contingent state-driven configurations of land and water usage that cut across traditional analytical boundaries between "the city" and peri-urban and rural areas in India.

Expanding urban development in Chennai has produced significant transformations in the configuration of land and water usage in and around the city. Chennai's population grew from 1,420,000 in 1951 to 8,653,521 in 2021.[10] Population estimates that include urbanized and suburban areas outside the city limits placed the population at over ten million in 2017. In order to keep pace with the corresponding rise in water needs for the city, the two major water organizations, the PWD and Metrowater, have engaged in a steady development of water sources and infrastructure. The city's major sources now consist of rain-fed reservoirs, groundwater, recycled waste water, and desalinated seawater. The state has made efforts to expand and diversify its sources of water, for instance by making rain water harvesting mandatory since 2002. Nevertheless, rain-fed reservoirs remain the primary source of Chennai's water supply. In 2016, Metrowater estimated that 65 percent of Chennai's water supply was provided by its reservoirs. The combination of Chennai's heavy dependence on rainfall for its water supply and the intense demands of urbanization have meant that reservoir supplies are inadequate sources of water supply for the city. In periods of drought, the city's supplies are placed in a crisis. The result has been that the city has increasingly relied on groundwater that is transported from rural and peri-urban areas (Butterworth et al. 2007; Janakarajan 2004).

In the context of extreme drought, with the failure of both monsoons in 2017, Tamil Nadu's water supply witnessed a severe crisis. Chennai's water reservoir levels had dropped below 13 percent by March, and water supplies across the state were drying up. The state government identified six hundred borewells across the state that would be used to supply drinking water for cities. Metrowater was designated to complete the diversion of water with a budget of Rs. 900 crores within four months.[11] This acute set of emergency measures in fact represented a much longer process of the state's diversion of water resources to meet city needs that had begun in earlier historical phases, in both the colonial and in the postindependence period. The emergence of such groundwater markets is not merely a natural offshoot of the shift toward privatization but a product of the intersection of state power and historically produced structures of political economy that precede recent decades of reform. In an important research study of irrigation law in Tamil Nadu, Carolin Arul (2008) has shown that the harnessing of irrigation water for Madras's water supply needs in fact stems back to colonial legal and state frameworks. In the early twentieth century, the colonial state would at various periods order the stoppage of irrigation purposes in order to ensure the supply of water to Madras (Arul 2008, 142). Such historical practices continued in the early decades of developmental activity in the postindependence period.

The expansion of water sources to meet Chennai's water needs gradually produced forms of infrastructural development that have transformed land regimes in both the city and the state of Tamil Nadu. This was facilitated by the strong authority that the state has over water resources. Land acquisition was historically always a dimension of the PWD's authority. The workplace code for employees specifically noted that "there is no objection to local officers negotiating with the owners of land with the object of coming to an amicable agreement" when necessary for the construction or management of water-related infrastructure through the legal framework of the 1894 Land Acquisition Act (GTN 1986, 64). The PWD also held the lease of land for the administration of water sources such as canals, drains, and channels (66). Meanwhile, in conjunction with the state's command approach to agricultural development in the first decades of independence, the PWD also exercised control of the water supply with "complete control over the larger works of irrigation" (GM 1958).

In the early decades of independence, the transfer of irrigation rights to serve the city's water supply began with the expansion of reservoirs designated to serve the growing urban population. The city's single major source of water from a rain-fed reservoir, the Poondi Reservoir, which was constructed in 1944, was expanded to include the Redhills Reservoir and Cholavaram Tank (see map 4.1). Irrigation rights from Cholaravam Lake and Redhills Lake were transferred for the city's supply in 1962 (Anbarasan 2010, 29). In the period 1966–69, the United Nations Development Programme (UNDP) conducted a series of studies that would first identify and recommend the usage of groundwater aquifers to meet city needs (31).

In an acknowledgment of the city's growing reliance on groundwater extraction in the following decades, a UNDP report noted that despite growing water needs, "fortunately, groundwater resources are proving to be a better than hoped for potential" (UNDP 1985, A-1). The report outlined the framework developed in conjunction with governmental proposals that would become the blueprint for the intensified extraction of water resources from rural to city consumption in the postliberalization period. The report specifically recommended the "purchase of irrigation water as a backup source of supply" and noted that the "Water Resources Planners report of May 16, 1985 outlines a proposal to call for farmers to forego the December 15–April 15 agricultural crop in the disaster 'double red code years'" (A-2). The agency then recommended the developmental framework that would underpin the establishment of water well fields that were not yet under the purview of Metrowater. As the report put it, "The mechanics for accomplishing the exchange of irrigation water for use by the city would be visualized as follows: A strategic reserve well field would be set up by legal description and legislation." The report identified the "Poondi-Tamarapakkam" well field as "the logical choice," given its proximity to Poondi Reservoir (A-2). The well field model of water supply for Chennai would later expand to include additional well fields and formed the underpinning of the underlying extraction of water from peri-urban and rural areas for city consumption.

The architecture of this planning process reveals two critical facets regarding the structuring of water markets in Tamil Nadu. First, the planning report underlines some of the historical continuities between the state-led developmental model associated with the early decades of Indian independence (in

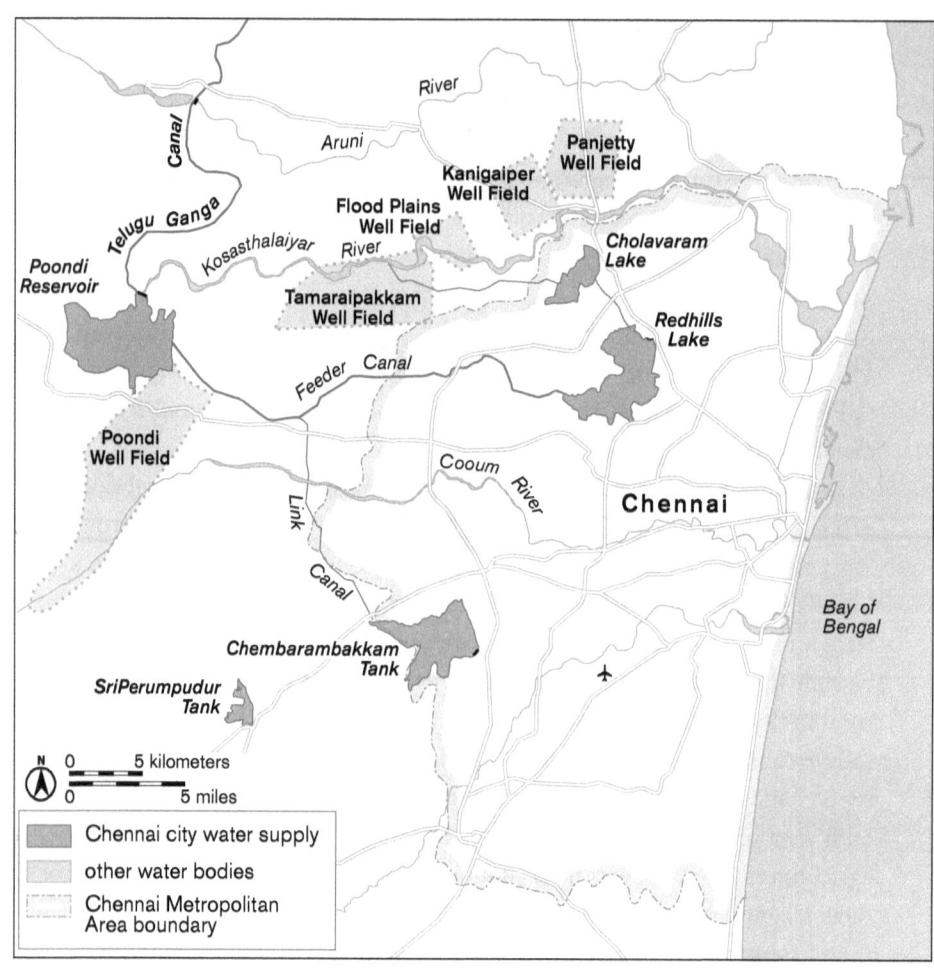

MAP 4.1. Chennai's Water Supply System, showing key sources of water that supply residents of the city of Chennai. The map draws on author's research and data from Chennai's major water utility (Metrowater).

conjunction with global developmental norms of the time) and recent trends in the postliberalization period. As we have seen with the effects of underlying structures of political economy on interstate water sharing between Karnataka and Tamil Nadu in the previous chapter, the underlying model of rapid, extractive development continues to shape the management of water resources. Second, water markets that now shape the distribution of water resources in the state have not emerged through natural rhythms of supply and demand but have been structured in significant ways by state practices and have continued to consolidate the centralization of state control over water resources.

Consider, for instance, how the report's call for "legal description and legislation" unfolded through legal and institutional reforms in the state. The report noted that the "city must, as a minimum, have control over i) drilling of new wells and ii) undesirable changes in cropping patterns" (UNDP 1985, A-3). The report then concluded with a broader recommendation for reform that would expand the authority of Metrowater, the bureaucratic institution that had now replaced the PWD in the management of Chennai's water supply. As the report stated,

> In order to introduce conjunctive use of water, the best course of action is to promulgate an ordinance which is necessitated to fulfill the intended functions of MMWSSB [Metrowater], i.e., to provide sufficient supply to water to cater to the needs of the ever and fast growing city of Madras.
>
> If conjunctive use and recharge of water is to continue on a long term basis it is possible, if the State Government is willing, to enact a bill to regulate and control extraction and use of ground-water in any notified area. Provisions for such a bill have been suggested in the model bill circulated by the Central Government.
>
> The State Government if it so desires, could also, extend the area of jurisdiction for the MMWSSB for certain limited functions and powers. (A-6)

Indeed, the Chennai Metropolitan Area Ground Water (Regulation) Act was enacted in 1987. The enactment of such regulatory legislation, in practice, has contradictory implications. Regulatory regimes are in fact necessary to manage the overexploitation of groundwater. In the case of Chennai, for instance, regulation of groundwater within the Chennai area was necessary to prevent the commercialized overexploitation of water through private markets. Metrowater was able to curb the commercial extraction of groundwater within the metropolitan Chennai area by stopping the issuance of permits for the extraction and sale of groundwater (PC 2007, 26). The result was the recovery of aquifers in southeast Chennai (the Thiruvanmayur Aquifer) and North Chennai that were being depleted by the commercial sale of groundwater. However, the UNDP report also illustrates the ways in which regulatory legislation has hidden links to developmental structures that are built on the political and socioeconomic power of cities in ways that reproduce centralized state authority through city-centric models of urban governance.

In the post-1990s period, while global, national, and state governmental policies and rhetoric promoted decentralization, regulatory state legislation was being melded with the centralization of state control. The political and economic dominance of the city of Chennai in relation to the surrounding urban and rural communities was encoded in dual legislation enacted for the state's authority over groundwater resources. While the 1987 bill was reworked for the Chennai metropolitan area in 2002, the remainder of the state's groundwater resources was placed under the purview of separate legislation. A parallel, stringent bill invested the government with the "power to develop, control, regulate and administer the groundwater in the State."[12] As with the Chennai metropolitan area bill, the Tamil Nadu Groundwater (Development and Management) Act, 2003, developed a regulatory framework based on a strict system of licensing and permits, and the state government placed restrictions on the hours of operation of pumps.

In practice, the bifurcated nature of this legislation both reflected and facilitated the extractive relationship between the city and neighboring towns and villages. While the expansion of Metrowater's regulatory powers has been effective within the metropolitan areas, growing water needs in Chennai have meant that Metrowater has continually expanded its own direct use of well fields in the metropolitan area as well as its reliance on groundwater supplies from peri-urban and rural areas. For example, in 1983–86, Metrowater had begun to expand its well fields, and the Thiruvanmayur Aquifer itself was taken over by Metrowater from the Tamil Nadu Water Supply and Drainage Board (TWAD), the organization that governs rural drinking water supplies (Anbarasan 2010, 31). The 1987 act thus implicitly encoded the expanding power of Metrowater and the primacy of city drinking water needs in addition to providing a needed regulatory system. The fractured legislation that separated out groundwater regulation in Chennai and the rest of Tamil Nadu further reflected and encoded this underlying imbalance in the regulatory system. Since Metrowater is not a governing authority accountable for the rest of the state's urban and rural groundwater resources, it is able to expand its reliance on rural water markets without any corresponding institutional accountability. The practical effect of this relationship has been that the groundwater market has continued to expand, and water is often pumped continuously over a twenty-four-hour period (interview, director of Centre for Water Resources, Anna University, August 16, 2016). Since the enforcement of groundwater legislation is itself structured by state power and underlying inequalities between the city and surrounding areas, it is unsurprising that the rules of regulatory structures for the rest of the state of Tamil Nadu have remained unimplemented.

The divided regulatory legal mechanisms facilitate the state's gradual redistribution of water resources from rural to urban metropolitan citizens. Meanwhile, the state government did not fully implement the 2003 act by framing specific rules and regulations, allowing it to take the form of a regulatory shell that would enable the continued extraction of groundwater.[13] As a Tamil Nadu government report would note, "In times of extreme drought condition, if the city is in need of water to be transported form distance [sic] sources, the Government may have to take a policy decision to suspend the irrigation rights (of course paying compensation for crop losses if any)" (GTN 2000, 9). The result is that the regulatory state enforces an extractive configuration of water (and land) usage that both builds on and produces unequal political-economic structures.

The creation of new regulatory state practices is part of a systematic process of postliberalization institutional reforms that were implemented through the major World Bank–funded Water Resources Consolidation Project. In addition to the institutional restructuring that I discussed earlier, a key dimension of this project was the focus on the mapping and management of groundwater and the implementation of an adequate institutional process for land acquisition and rehabilitation for people displaced by water-related infrastructure projects. The state's mapping of groundwater resources has increasingly become a critical dimension for the management of water resources. The WRCP established the State Ground and Surface Water Resources Data Centre, with improved technological capabilities that now provide monthly monitoring of control wells. However, as a reflection of the stratified institutional field, the monitoring of groundwater is also conducted by Metrowater and TWAD within Tamil Nadu as well as the central government's Central Ground Water Board. Given the increasing pressures of water scarcity, data collection on groundwater has become one of the most significant dimensions of both state planning and state power.[14]

The ability of the state to track groundwater resources in periods of crisis becomes one of the central means of extracting water for consumption, primarily for the Chennai metropolitan area. As early as the late 1990s, a government report would note that "in drought years the Chennai Metropolitan Water Supply System is exploiting groundwater [in the Chennai basin] to the maximum extent possible" and that the overexploitation had begun to produce seawater intrusion (PWD 1997). By 2004, after a period of drought produced by deficient rainfall, the Data Centre would report that "almost in the entire city [the] water level has gone down considerably when compared with water level of January 1994 ... [indicating] enormous pumping of groundwater during the last decade" (CE 2015). By 2017, after a period of severe water scarcity that was produced by another failed monsoon, the exploitation of groundwater had reached a severe crisis in the state.[15] The state's mapping and regulatory control of groundwater resources will thus continue to remain a central site for the exercise of state power.[16]

However, despite this sustained process of water mapping by the state, the institutional disjunctures that I have addressed shape the extent to which this state endeavor translates into sustainable water policies. We have already seen that the PWD's Water Resources Organisation often cannot effectively manage water supplies in sustainable ways in the face of both state governmental policies that continue to promote urban development in the context of lucrative real estate deals and a metropolitan city-centered model of liberalization. For instance, most policy decisions continue to be based on land usage and land cover data rather than on groundwater storage (Chinnasamy and Agoramoorthy 2015, 2140). Regulatory state authority of land and water are, as we have seen, implemented by separate sets of bureaucratic institutions. This account of institutional cleavages that have fragmented regulatory mechanisms is more than a mere story about the dysfunctions produced by institutional fragmentation. Institutional reforms that have sought to produce rationalized efficiency and decentralization have produced a differentiated bureaucratic field that mirrors broader political-economic processes of restructuring. There is, in effect, a redistribution of institutional power that encodes the inequities that shape the political economy of India's liberalizing state.

The ascendancy of India's metropolitan cities within India's liberalizing economy has meant that utilities serving metropolitan areas have also grown in power. While the restructuring of the Public Works Department occurred in the late 1970s, Metrowater's institutional power has continued to grow in relation to both the PWD's Water Resources Organisation and TWAD. The spatial aesthetics of the PWD's irrigation branch and Metrowater in many ways embody the shifting relationship. PWD's irrigation branch is housed in the imposing colonial building that embodies the historical legacy of its political and economic power. Yet the building is sparsely occupied, without any of the technological upgrades that are used to brand India's new economy. In contrast, Metrowater's smaller complex has the visual markers of this new economy. Flat-screen televisions are lodged over elevators displaying the utility's technological upgrades at its reservoirs and desalination plants.

This, of course, does not mean that the PWD does not have its own sites of power within the water bureaucracy. The newly reorganized Water Resources Organisation, which manages the state's water sources, has had significant technological upgrades, particularly in relation to the detailed mapping and data collection of groundwater resources. However, the dynamics between the PWD and Metrowater were succinctly captured by the engineer overseeing the Chembarambakkam Reservoir. Standing at the top of the supply tower, we could see the brand-new black pipeline that carried water to Chennai. Near the pipeline were two white buildings owned and run by Metrowater. Pointing to a smaller, shabbier building owned by the PWD, he recounted, "Every year, Metrowater comes and whitewashes the buildings, but they never do the PWD building" (interview, January 19, 2017).

While interagency cooperation is crucial for the management of Tamil Nadu's water supply, the steep competitive strains between urban and rural water users in the context of water scarcity have been reproduced within the institutional divisions of the water bureaucracy. The management of groundwater, for instance, falls under the purview of numerous organizations, including TWAD, Metrowater, PWD, the Directorate of Rural Development, and the Agriculture and Farmers Welfare Department. Internal governmental reports point to the lack of integration between these departments. For instance, the

World Bank–funded consultancy report called attention to the obstacles to shifting irrigation resources to drinking water supply needs. Arguing that the "water resources organization (PWD) enjoyed a strong lobby for irrigation needs," the consultancy firm went on to note that "such absence of coordination between the departments results in water not being allocated according to the declared Water Policy Priorities. Requests for the provision for drinking water from new storage projects, earmarked for irrigation by WRO, are usually denied" (WRCP 2001, 23). An internal governmental review would echo this perspective on competing institutional agendas, noting, for example, that "a plethora of agencies are involved in watershed management of the catchments," producing a kind of fragmentation in which "with [the] formation of smaller districts [the] absence of a pro-active leadership and central authority for coordinating the activities of the various agencies and departments and for focusing on effective water resources management is acutely felt" (GTN 2003, 62). On an everyday level, one assistant engineer noted that the sharing of data would often become a source of contention between different wings of the water bureaucracy. Given the scarcity of water sources and the intense competition between departments representing different water users and consumers, scientific data on the availability of existing water supplies becomes a critical site for control and contestation within the stratified water bureaucracy.

The nature of such contestation is shaped by both political considerations and structures of political economy. For instance, the deepening inequalities between Chennai and the rural and smaller urban areas have weakened the institutional power of TWAD, the agency in command of rural drinking water supply. As Govind Gopakumar has argued, "Unlike Metrowater, the TWAD Board has been unable to maintain a revenue surplus as a result of the inability of many small rural and urban bodies to pay their bills. The irregular flow of revenue has directly threatened the existence of the TWAD Board. The institutional robustness of Metrowater and the corresponding weakness of the TWAD Board have reinforced the distinction between the availability of water supply in Chennai and its periphery" (2012, 118).

In the past, electoral considerations meant that particular rural constituencies could hold state officials accountable. However, the rising importance of urban development and a growing urban middle-class dominance of public spheres of communication have also produced a shift toward the political and urban power of city centers such as Chennai. The shifting relationships of power within the water bureaucracy are not a story of declining state power but a shifting of power between state structures and institutions.

The regulatory frameworks of water management in Tamil Nadu have been shaped in significant ways by underlying structures of inequality that have, in practice, transformed regulatory practices into an extractive relationship both between urban and rural communities and within these communities. Such structural contradictions rupture the state's regulatory framework, as they exceed the state's ability to manage this extractive relationship. The overexploitation of groundwater sources in Chennai has meant an increasing reliance on the supply of groundwater from rural areas in ways that contradict Tamil Nadu's formal legal regulations. The result is that the state's regulatory framework itself has been placed in a conflicted state of paralysis. In 2013, in recognition of both the gap between the formal framework of the law and the actual exploitation of groundwater and the reliance of this extraction for water supplies, the Tamil Nadu government repealed the 2003 groundwater act.[17] A year later, the government attempted to pass new ordinances both placing regulatory limits on new construction outside the metropolitan Chennai area that would impact groundwater and banning the extraction of water by packaged drinking water industries from groundwater blocks with either an overexploited or a critical status.[18] The ban on packaged water units was itself an attempted retroactive regulatory correction, as a 2012–13 report by the comptroller and auditor general of India had already documented the unregulated exploitation of Tamil Nadu's groundwater by the packaged water industry due to the absence of adequate state regulation.[19] As one media report noted, such units had to gain a No Objection Certificate from the state and then apply for a license from the Bureau of Indian Standards. Yet while state water authorities had only issued such certificates to 2 of 49 units that had applied for licenses, 440 units had received licenses. As with the institutional fissures with organizations overseeing land usage, the water bureaucracy, with the knowledge of the deep strains on groundwater, was not able to effectively wield regulatory power in the face of more powerful bureaucratic entities. Further, complicating these regulatory failures, the packaging industry filed a lawsuit challenging the new ordinance by capitalizing on the state's own regulatory failures. The industry argued that the state's own repeal of its 2003 act without implementing it meant that the state had no legislative authority to regulate groundwater.[20]

Regulatory state frameworks have thus inadvertently been transformed into mechanisms for the extraction of water supplies to serve the Chennai metropolitan area. In this context, the structure of Chennai's water supplies is a fraught story of conflict over land and water that cannot be understood through a methodological or analytical lens that reifies the territorial boundaries of the city. The contours of this water market are structured by conscious practices of state intervention and the withdrawal of state action in this management of land and water usage. An understanding of the making of water markets in the city and state thus necessitate an analysis of the ways in which state power reconfigures land and water in ways that build on historically contingent political-economic structures of inequality and city-centric developmental extraction in the postliberalization period. Such practices are shaped by domestic political considerations that are in turn contingent on relationships of power between socioeconomic groups.

Land Usage, Water Markets, and the Reconstitution of Public Welfare

In the postliberalization period, the production of water markets through the extraction of groundwater in rural and peri-urban Tamil Nadu occurs along two major pathways that center on both state intervention and the withdrawal of state action. The state's focus on groundwater extraction draws on a long history of state-led agricultural development in India that produced a major turn toward tube well irrigation. Tamil Nadu is one of the largest producers of agricultural products in India (Chinnasamy and Agoramoorthy 2015), and the state's reliance on groundwater for irrigation has led to a full utilization of water supplies for irrigation and also spurred a corresponding shift from noncommercial to commercial crops. It is worth noting that while there is now a systematic transfer of water resources to the Chennai metropolitan region, irrigation for agriculture still represents the largest portion of water usage in the state (see table 4.2). What is changing, then, in postliberalization is not the state's command of water resources for particular economic activities but the state's priorities. The postliberalization period is marked by a shift in state priorities toward urban-led development and the drinking and industrial water needs for consumers and investors in the Chennai metropolitan area. This shift, as we have seen, builds on both colonial and postindependence trends where the state has actively shaped

TABLE 4.2. Sectoral water demand, Tamil Nadu

Water demand by sector	1994 (MCM)	2001 (MCM)	2010 (MCM)	2020 (MCM)
Irrigation sector	2,066.0	49,978	43,220	49,850
Domestic sector	181.8	2,222	1,000	1,200
Industries	86.23	1,555	1,500	1,700

MCM = million cubic meters. Sources: PWD 1994, 148; GTN 2003, 47; 2010 and 2020 figures are estimates, cited in Suresh 2021, 7.

the transfer of irrigation water to supply the city's needs when needed. For instance, pumping from peri-urban villages started as early as 1965 (Janakarajan et al. 2007, 54). What has changed is the intensification and systemic nature of this transfer and the reforms of regulatory state practices that enable this transfer.

Consider, for example, the impact of Metrowater's new legal powers over water resources. The expansion of Metrowater's powers has enabled the utility to directly purchase water rights from farmers. The utility, of course, operates under pressures of its own, as it is faced with the task of meeting rising water needs in an expanding metropolitan area in a city and state that experiences chronic water scarcity. One senior engineering executive explained to me that in the 2002 drought, Metrowater had to "convince farmers" to supply water for Chennai's drinking water needs and had a Rs. 1 crore daily expense when water was supplied purely by lorries (interview, August 17, 2016). Or, to take another example, NGO project staff working in peri-urban areas "were informed that the officials invoke an emotional argument while searching for water sellers: *that if you cannot supply water to your own people in Chennai, how can we ask water for our farmers from Karnataka?*" (Janakarajan 2004, 10) In this context, Metrowater becomes the arm of the state that draws on both financial incentives and the affective dimensions of ethnicized citizenship that have been intensified in the context of interstate conflicts over water.

Metrowater's role in structuring water markets in ways that produce a transfer of water from rural and peri-urban to city needs has been reinforced by other state structures in Tamil Nadu in the postliberalization period. Madras High Court decisions have shown preference given to supply drinking water to the city and the state's purchase of water rights that enable the transfer of irrigation tanks to serve city water supply needs (Arul 2008). A significant example of this is evident in Carolin Arul's discussion of the New Veeranam Extension Project, designed to transfer water from Veeranam Lake in Cuddalore District for Chennai's water supply. The scheme, first conceived in 1969, began to take shape in earnest only in the midst of India's liberalized developmental expansion in the 1990s and was finally commissioned by the AIADMK government in 2002.[21] Arul's research shows that both political leaders and the courts intervened (including a personal visit by the chief minister in 2004) to preserve the rights of the state to divert water in the face of farmer resistance to the project. A court case that "protested the hardship to agriculturalists and preferential treatment for Chennai metro residents" and included a "a flood of letters from farmers including some signatures with blood" (Arul 2008, 232) was dismissed with the court simply asking the government to explain the project to the farmers.

What is central to an understanding of the postliberalization state is that the government, in this context, is not merely mediating between competing water users or legal parties. Rather, the combination of legal judgments, policy frameworks, and executive decisions taken together reinforce the state's rights over water resources rather than those of water users who may have had long-standing rights based on use. Conceptions of public welfare and public trust become the means for a recentralization of state authority. As in the colonial and twentieth-century developmental periods, the state asserts claims of protecting the "common good" and representing the public interest by asserting its authority over water resources. It is the state's definition of public welfare that shapes the structuring of water markets. In the case of the New Veeranam project, the state's response to the court case was the 1994 Water Policy of Tamil Nadu, which prioritizes drinking water needs and which is in accordance with the framework of India's National Water Policy. As one assistant engineer at the PWD put it, "The water needs are in agriculture, but we are told to give priority to drinking water" (interview, January 19, 2017).

This practical and political determination of policy priorities represents a process of restructuring that is more than a reflection of long histories of unequal development. State policies and the distribution of resources in India have long been shaped by the interests of dominant social groups in ways that have in turn intensified inequalities that have undergirded formal citizenship rights. The dynamics of the postliberalization state in this context do not represent a retreat from the theories and ideals of India's version of social welfare norms. The regulatory regimes that are set up in conjunction with policies of reform produce a framework for the state's reassertion of its long-standing authority over public social welfare. In this context, the welfarist dimensions of the state are not reduced; rather, they are redistributed in line with the new policy goals of a liberalizing state.

This authority of the state over water resources has been backed by the Supreme Court, where a 1997 decision reinforced the conception of water as a public trust, where the "state as a trustee is under a legal duty to protect the natural resources. These resources meant for public use cannot be converted into private ownership" (Cullet 2009, 43). However, the idea of the public trust has been shaped by distinctive hierarchical and spatialized conceptions of the public sphere in the postliberalization era. The public good is increasingly identified with specific, dominant representations of metropolitan middle-class citizens (Fernandes 2006) and the new model of city-based economic growth, which has been characteristic of the postliberalization period and is now embodied in governmental programs such as the Smart Cities Mission. The irrigation-driven strategy of the early decades of developmentalism that was linked to food security needs has now been replaced by a form of growth that is largely driven by new economy sectors such as the services sector and IT, which are concentrated in metropolitan cities and their surrounding urbanizing areas.[22] For example, according to the Ministry of Finance's *Economic Survey*, the services sector "contributed almost 66.1% of its gross added value growth in 2015–2016," making it a crucial foreign exchange earner (MF 2016). Given that water consumption is much higher in cities than in rural areas, such patterns deepen inequalities between rural and urban areas. While irrigation remains the primary sector in terms of overall water consumption, the growing significance of urban-led growth is in the process of restructuring the distribution of water resources in significant ways. Shifts from the developmental state's promotion of the rapid expansion of agricultural productivity to address food security in the early decades to an increasingly city-based state strategy of economic growth have intensified the competition for water resources between different sets of users, ranging from industries to farmers to various social groups in urban and rural areas (Ballabh 2008; Joy et al. 2008).[23]

The socioeconomic strains produced by the effects of this reorientation are illustrated by the state's reactions to farmer suicides brought on by financial distress and severe drought in Tamil Nadu in 2017. In response to a public interest litigation suit filed by an NGO in Tamil Nadu (Tamil Nadu Centre for Public Interest Litigation), the Supreme Court ordered the Tamil Nadu government to address the plight of farmers and to provide a reply to the court within two weeks. In their rejoinder, the Court bench noted,

> The state stands on the position of a loco parentis to the citizens and when there are so many deaths of farmers in the state of Tamil Nadu, it becomes obligatory on the part of the state to express concern and sensitiveness to do the needful and not allow the impecunious and poverty stricken farmers to resign to their fate or leave the downtrodden and the poor to yield to the idea of fatalism.... *The concept is alien in the welfare state and social justice which is required to be translated into a democratic body polity* [emphasis added]. As is manifested from the assertions and the grievances, deaths are due to famine and other natural causes and also due to immense financial problem[s]. The state, as the guardian, is required to see how to solve these problems or to meet the problems by taking curative measures treating it as a natural disaster. Silence is not the answer.[24]

The rhetoric of the bench, while laced with paternalistic conceptions both of the state-citizen relationship and of farmers, provides an acute statement on the need for the preservation of the responsibility of the welfare state. At one level, this response illustrates the contested nature of the Indian state and the potential for political and social pressure within the contours of democratic state institutions. However, at another level, this intervention reflects an institutional pattern in which the Supreme Court once again exceeds its traditional purview of power and authority because of the failures or lack of executive governmental action. As with the case of interstate conflicts, failure of action by both the central and local state governments prompted the Supreme Court to intervene in a policy arena that should traditionally fall within the executive branch of state authority. The Supreme Court intervention in this case (which also occurred after the Madras High Court refused the petitioner's plea) reflects the underlying recasting of the welfare state, which in theory "is required to be translated into a democratic body polity" but in practice has been reoriented to serve new state norms of welfare in the postliberalization period, which in this case prioritize city over rural needs.

The production of such state priorities is not adequately understood purely through stereotypical conceptions of bureaucratic indifference or corruption. For instance, the identification of drinking water needs as a priority in national and state governmental policies is a goal that is, in theory, fully in keeping with and a necessary dimension of an inclusive conception of the welfare state. Furthermore, the assertive moves of a water utility are fully in keeping with the bureaucratic objectives of providing water for Chennai's population. Water shortages in Chennai are, of course, a real crisis. What is at stake is an understanding of how economic policies in the postliberalization period have redrawn the regulatory boundaries of the welfarist dimensions of the state in line with the investment-driven urbanized centers of development and progress and the corresponding models of water markets that serve these centers. The state has in effect been actively shaping water markets in and for the city of Chennai.

A second dimension of the state's role is alluded to in the Supreme Court's admonition, "Silence is not the answer." The postliberalization state also structures water markets through an absence of action or intervention. This lack of institutional capacity is not identical to formal policies of privatization that curtail the role or power of the state in order to draw in the private sector. Rather, private markets emerge when the state either fails or chooses not to intervene without necessarily abandoning any formal authority or power. Consider, for instance, the expanding groundwater market, which, as we have seen, has increasingly become a primary source of water for both domestic and industrial users in Chennai. The anecdote that I began this chapter with points to the ways in which the state's gradual withdrawal of water resources for irrigation has allowed urbanization to take place.[25] The result is, paradoxically, that the lack of adequate supply of water for agriculture furthers the impetus of farmers to resort to selling groundwater. As a Government of Tamil Nadu report notes, "Many farmers have reported that mainly dwindling water supplies from the wells and increased labour problems both in terms of wages and availability constrained irrigated crop production. Further increased cost of inputs compared to output prices discouraged irrigating several crops. Hence farmers were forced to sell the water after meeting their requirement for standing coconut and other crops" (GTN 2003, 139).

In addition to declining water supplies leading to the sale of water, other groups of farmers must also rely on the purchase of water to supplement the exhaustion of well tanks (GTN 2003, 139–40). Other forms of state practices have also inadvertently contributed to the emergence of water markets. The populist agendas of Tamil Nadu's electoral politics have included the provision of heavily subsidized electricity to farmers. This has meant that pumping of groundwater has been a financially profitable endeavor for landowners with rights over groundwater (Janakarajan et al. 2007; Packialakshmi, Ambujam, and Nelliyat 2011). While, as I have noted, there are restrictions on the hours of pumping, the state's nonimplementation of such regulations becomes a de facto method of enabling groundwater markets to continue to supply water for urban needs. The economic effects of such markets are themselves contradictory. In some cases, marginal farmers may benefit from the sale of groundwater, while landless laborers stand to suffer the most, as they lose employment with the decline of agriculture (Packialakshmi 2012). Meanwhile, the regulatory system itself produces contradictory effects. As one government report notes, farmers complained that when they received government loans to dig wells, the delays they experienced in getting clearance certificates placed them at a disadvantage, since for wealthier farmers, "the certificate need not be obtained when people dug wells with their own money. Due to this the wells already dug by farmers after obtaining the clearance for minimum spacing get affected and causes reduction of yield in the wells" (GWB 1992). Thus, the investment in groundwater markets for farmers is also a risky venture, particularly for less well-off farmers, as the nature of groundwater is fluid and the extraction of wells in one area has a significant impact on neighboring wells. In this context, the state's early attempt at regulating groundwater extraction in the 1990s had an inadvertent detrimental effect on less privileged farmers who are dependent on government loans.

The emergence of water markets is a product of a diverse set of state policies and absence of action in the context of a model of development that continues to place needs on the supply of water for both domestic and industrial consumption in and around Chennai. While internal institutional conflicts over the supply for rural versus urban areas are often cast as a conflict between agricultural irrigation needs and drinking water supplies, Metrowater's purchase of water from farmers is also designed to serve industrial needs (Ruet, Gambiez, and Lacour 2007). The result is the creation of a multitude of practices that form an informal groundwater market on the periphery of the city. In South Chennai's IT corridor, twelve hundred tankers provide water to this peri-urban area per day (Packialakshmi, Ambujam, and Nelliyat 2011, 427), with deleterious effects for agriculture in neighboring villages that the water was being extracted from. Indeed, the continual movement of water tankers is a common sight on the streets of Chennai and is a continual

EXTRACTION, INEQUALITY, AND BUREAUCRACY 173

FIGURE 4.2. Water Tanker in Chennai, showing one of the private water tankers that routinely transport water from rural and peri-urban areas to the city of Chennai. Chennai's businesses, residents, and water utility (Metrowater) purchase the water from private dealers.

visual reminder of the daily extraction of water resources for urban needs in and around the city (see figure 4.2). The overextraction of groundwater has had further ecological impacts, as it has resulted in seawater intrusion that has further jeopardized water sources for the city (interview, chief engineer of irrigation, PWD, January 11, 2017).

The systemic extraction of rural groundwater for urban needs points to the ways in which the regulatory state is shaped by deeper relationships of power. At one level, state practices shape markets through inaction—that is, through forms of regulatory failure and institutional incapacities, as well as the withdrawal of action. Water markets in this context are not the creation of reform-driven models of privatization but the result of an accumulation of informalized practices that fill the void produced by state incapacities or intentional inaction. The result is that both state action and state inaction have redistributive effects that undergird the global and state languages of technocratic efficiency and management. More significantly, the nature of the regulatory state in this sector is such that the burgeoning formal and informal private water markets are a product of state planning. That is, the intentional *withdrawal* of state action—in this instance by not enforcing existing groundwater legislation—is an interventionist state strategy of managing and consolidating the dominance of urban-led development, which has intensified in the postliberalization period.

State Power and the Question of Privatization in Chennai

The postliberalization period in India is generally associated with both the rhetorical and policy shifts that have foregrounded the private sector and the need for private investment in various sectors of the economy that were once the purview of public sector control. In the context of Chennai, as we have seen, the effects of this model on water resources and infrastructure have been an indirect one embodied in the intensification of urban development and the corresponding shifts of the usage of land and water. Such developmental models are concrete examples of new business-state relations that shape the political economy of liberalizing India (Jaffrelot, Kohli, and Murali 2019). However, the dominant global model that encourages infrastructure funded by private capital has not significantly shaped the construction and management of water infrastructure either in Chennai or in rural areas in Tamil Nadu.

This necessitates a rethinking of public debates over water sector reforms in India, which often splinter into political positions in opposition to or in support of privatization that do not capture the complexities of state power. Consider, for instance, some of the broad patterns of private and public control over water resources and infrastructure. Research on changes in the control over water has demonstrated that there are some cases in India that can serve as examples of straightforward forms of privatization. Examples of overt forms of privatization include the privatization of rivers in India, such as the privatization of a river in Chhattisgarh through the lease of a stretch of river to a company (Cullet 2009, 48). Or, to take another well-known example, the rapid expansion of the soft drink and bottled water industry produced a high-profile court battle to ban Coca-Cola and Pepsi products in Kerala (Aiyer 2008).

A closer analysis of patterns of privatization shows a more complex configuration of the relationship between the public and private control of water and water-related infrastructure. Consider, for instance, the case of public-private partnerships, one of the key dimensions of the new global-national model for water governance. Trends do show an increase in the establishment of water sector PPP projects in India. By 2011, the World Bank estimated that there was a gradual growth of such projects, with five million in urban areas receiving water from institutional arrangements involving private sector participation (Swaroop 2011, 6). Patterns of privatization are discernible in the water sector in India but do not dislodge conventional forms of state bureaucratic authority. The state, in effect, remains the central actor that controls water resources. While international organizations such as the World Bank have provided significant funding for water-related projects in the state, such projects have been implemented and managed through state institutions.[26] Reforms have taken the form of institutional restructuring and the subcontracting of projects in order to streamline these institutions. However, while such restructuring has reworked the relationships between institutions, the state has maintained clear control over water resources. In times of crises such as floods and drought, the state government's focus has been on pressing claims for relief and compensation from the central government. The role of private capital in this context has been focused on smaller urban localities.

The model of privatization that has been implemented as part of reforms in Tamil Nadu's water sector has been one that has focused more on internal, workplace restructuring within the water bureaucracy. Such reforms have unfolded along the familiar lines of the reorganization of management and the streamlining of the staff of both Metrowater and the Water Resources Wing of the PWD. In the case of Metrowater, the utility has systematically engaged in a reduction of its staff, even as the area of coverage under the utility has expanded. While the utility had shrunk from 7,400 to 2,060 employees by 2016, it had added forty-two urban local bodies covered by its water supply in 2011 (interview, August 17, 2016). The restructuring was accompanied by practices of subcontracting of both planning and infrastructure construction contracts to external consultants. Meanwhile, in the case of PWD, staff reductions have occurred through the maintenance of vacancies rather than more politically charged processes of retrenchment. While senior ranks of the organization have been maintained, vacancies for junior level posts are either left vacant or hired on temporary project-related contracts rather than in permanent positions (interview with assistant executive engineer, PWD, January 18, 2017). For example, World Bank–supported reforms that reorganized the PWD along the lines of basin river management were well received by senior employees because the reorganization expanded the number of chief engineers (the senior-most rank) within Tamil Nadu; each river basin thus now has its own chief engineer (interview with chief engineer of irrigation, PWD, January 11, 2017).

In contrast, the major global privatization principle for the water sector—the prescribed move toward cost recovery through water metering—has been slow in its implementation. Consider, for instance, the ways in which the political difficulty of charging urban users for water consumption in Chennai has stunted such reforms. One of the major global norms that is put forth by global institutions such as the World Bank is the construction of water as an economic commodity. The World Bank has systematically promoted projects and policies that have required establishment of water meters and user fees as a way of rationalizing the use of water. However, eight years after this initiative was launched in Chennai, the utility was still trying to jump-start it by beginning to meter a set of commercial buildings.[27] Indeed, in my interviews with engineers at Metrowater, an overhaul of the metering system and the use of smart meters was still being presented as a major new initiative needed to manage consumption, particularly given low charges for water usage (interview, August 17, 2016). By the beginning of 2021, Metrowater was set to complete the installation of meters in all commercial establishments with plans to expand this to consumers in Chennai.[28]

While critics rightly point to the problems of global dominant discourses that commodify water (particularly for socioeconomically marginalized communities), in practice the lack of metering has also subsidized wealthier communities in the city. Without a systematic metering system, Metrowater has used the control of water supply, through control of the hours of piped water supply, as a means of managing consumption. One senior engineer at Metrowater, for instance, noted that given the low water charges, this was the only means the utility had for managing supply and consumption (interview, August 17, 2016). The pressures of scarce water supplies have been such that the utility keeps track of how water is being consumed through the specific monitoring of pipe supply to kitchens and to the rest of the household, as there are separate pipes for these two kinds of supply (interview with professor of civil engineering, Anna University, August 16, 2017). This has meant that water is supplied by Metrowater for two to three hours a day. While wealthier consumers can supplement this by purchasing private water, low-income communities rely heavily on supplies from the utility. In this case, while the formal commodification of water through metering has been forestalled, there has in fact been an informal commodification of water through the default reliance of wealthier consumers and businesses on private suppliers. This in turn leads to the prevalence of corrupt and organized networks colloquially known as the "water mafias." Privatization here unfolds through the limits of the water bureaucracy's ability to effectively regulate water through formal practices. The inadequacies of regulation open up the space for informal and formal water markets to emerge.

These practical microdynamics illustrate the complexities involved in the emergence of water markets. Privatization in this context is a subsidiary process in the reforms that have been carried out. The kinds of inequality that critics of privatization have been concerned with are shaped not by a simple transition from public to privatized goods but through a reworking of which public matters. For instance, the government provision of free water that is directed toward the benefit of relatively privileged consumers with access to piped water supplies and the ability to pay for water may inadvertently intensify both inequities of access to water and the skew of the distribution of public resources toward wealthier urban groups. Consider, for instance, the Aam Aadmi Party (AAP) political party's promise to provide free water in Delhi in the 2014 elections, which marked its first major electoral success. The provision of water in Delhi requires the long-distance transportation of water from groundwater sources in other locations, which could in effect serve to reinforce an extractive relationship in the name of equity. According to some estimates, as much as 70–80 percent of water subsidies do not reach the poor (Foster, Pattanayak, and Prokopy 2003; McKenzie and Ray 2009). On the other hand, the promise of 24/7 service delivery in exchange for user fees, one of the key features of the dominant model of water sector reforms, would also produce acute inequities for communities that either cannot afford to pay user fees or do not have access to piped water. Such questions of access are particularly significant given the ways in which caste structures access to water at the local level; low-caste communities may in effect not have adequate access to water, even if communal piped connections exist or are provided through infrastructural development. An adequate assessment of the impact of water sector reforms on such questions of equity complicates ideologically driven positions for or against privatization. In the backdrop of such nuances lies the fact that reforms have often intensified state centralization and intervention rather than practices of decentralization or participatory management. What then becomes of two of the key facets of such reforms—the principles of privatization and decentralization? Such principles in effect target both the less powerful segments of the water bureaucracy within the metropolitan city and less politically and economically powerful sites in rural and small-town India.

Private Capital, Reforms, and the Remaking of State Power in Small Towns and Rural Communities

Significant institutional and financial restructuring, which is conventionally associated with economic reforms, has largely focused on urban communities that are classified under the rubric of "urban local bodies" (ULBs).[29] Tamil Nadu has developed a financial model for infrastructure development that is often portrayed as both a national and global model for structuring public-private investment for ULBs. The Tamil Nadu Urban Infrastructure Financial Services Limited (TNUIFSL) has emerged as a highly successful fund manager that raises private funds for the Tamil Nadu Urban Development Fund (TNUDF). The TNUDF was established by the Government of Tamil Nadu in 1996 as "the first public-private partnership providing long term financing for civic infrastructure" (TNUDF 2016, 1). The fund was based on financial models advocated by the International Bank for Reconstruction and Development (Mahalingam, Devkar, and Kalidindi 2011). The TNUIFSL funds are provided as loans to urban local bodies for infrastructure development and have been a central means of restructuring local administrations in these urban areas. In contrast to other sectors of the economy, private capital has not had a significant interest in investment in water infrastructure. The complexities of managing and maintaining water-based infrastructure (including high costs and the length of time for the implementation of such projects) have made the water sector a less attractive option for private investors. TNUIFSL is thus in many ways a distinctive enterprise, as it includes water infrastructure in its lending program. However, the specific fund for water-based infrastructure, the Water and Sanitation Pooled Fund, is a trust owned and fully funded by the government (through grants or loans taken out by the government) (interview with managing director, TNUIFSL, January 12, 2017). As the managing director of the fund noted, there are significant challenges to raising private funds for water infrastructure, and the fund is a pooled fund because the funds are smaller.

The TNUIFSL in effect manages the disbursement of both private and governmental funds in order to enforce objective financial and planning standards without the intervention of politically oriented state agencies. Money from external international agencies is sent first to the Government of India, then disbursed to the Tamil Nadu government, and finally managed by the fund. The financial structure ensures that neither the central nor the state government directly spends the funds received (interview, January 12, 2017). As the managing director said, "We think of ULBs as corporations not government." This approach to ULBs is echoed in the funds planning approach, which is framed around a city corporate and business plan (TNUIFSL, n.d.). ULBs seeking a loan must develop a city corporate plan in order to demonstrate that they are able to illustrate long-term financial planning (interview, January 12, 2017). The fund approves loans only for local bodies that demonstrate financial viability. This strict approach has made the fund a highly successful financial enterprise. In its first sixteen years of operation, from 2002 to 2020, it has reported a "100% collection efficiency," making it a model that has now attracted international attention (TNUDF 2020; interview, January 12, 2017).

TNUIFSL's model of financing and urban infrastructure development is part of the larger set of economic reforms that have emphasized financial decentralization and have devolved funds to local governmental bodies. It is in this realm of weaker and smaller urban localities that we see the dominant national-global model of privatization and decentralization being implemented. However, there are, even in this context, limits to this implementation. Urban local bodies often have had limited resources and have had to resort to taking out loans from financial agencies. Moreover, while the establishment of financial models such as TNUIFSL was intended to create independent financial pools of funding without government support, in reality private investors have been wary of the risks involved in supporting both ULB infrastructure projects and water-related infrastructure in particular. Sonia Hoque has noted that the majority of urban infrastructure projects in ULBs "depend on subsidized funds from state governments and semi-public financial institutions that lend to ULBs relying on state government guarantees" (2012, 7). In 2015–16, close to 49 percent of the Tamil Nadu Urban Development Fund's financing came from the state and central governments. Financing for water-related projects has required governmental backing in order to mediate such risks (Venkatachalam 2005). TNUIFSL's financial model has worked because it has been backed by government guarantees as well as a significant credit line from the World Bank. Hence, the model does not represent a clear-cut case of a shift toward the privatization of the water sector.

Financial requirements of budgetary discipline are indeed imposed on urban local bodies in accordance with the norms of private financial investors. However, the investors are sheltered from financial risk by governmental protection. This form of privatization in fact does not represent a retreat of state support; rather, it represents a shift from the state support of local governments to the protection of financial capital. L. Krishnan has noted that TNUDF, the development institution that is managed by TNUIFSL, was specifically "designed to take urban infrastructure financing out of the realm of government budgetary allocations and regulations and instill it with a business orientation that would accelerate financing decisions and encourage innovation" (2007, 238).[30] As Krishnan, who was former special secretary to the Government of Tamil Nadu, further notes, this was in large part due to a lack of state resources for urban infrastructure. In 2001, the state of Tamil Nadu needed an estimated $2 billion for infrastructure for ULBs, with a significant portion of this needed for water infrastructure (242). Within the broader structural relationship of inequality between the Chennai metropolitan area and rural communities, ULBs have tended to suffer from significant deficiencies in water infrastructural development, including the lack of the adequate provision of drinking water supply (Harriss-White 2016, 4). However, while the intention of this program was to address such infrastructural problems through private financing, in practice the model resorted to replacing the state support of local governments with the state support of private capital.

TNUIFSL's model of public-private funding has both similarities to and differences from dominant global models that have stressed financial soundness over questions of citizenship access and equity. In the case of water-related infrastructure, the fund adopts global norms of enforcing water tariffs. In contrast to Chennai, water supplies and infrastructure in ULBs that receive funds require the acceptance of user tariffs. However, while there is a one-time connection fee for households, monthly tariffs are determined on a graded system based on landownership. According to the managing director, the fund has the objective of providing "equitable and continuous supply of water" and ensures that water infrastructure projects take a holistic approach that encompasses all connections from the water source to the user. While Metrowater's inability to institute tariffs has meant that wealthier households have benefited from state-subsidized water resources, the TNUIFSL has attempted to institute an equity-based system that deviates from global models of water tariffs in ways that seek to address socioeconomic inequality (interview, January 12, 2017).

However, the business-oriented model of financial viability also produces other kinds of inequity for other rural and urban localities. The vast majority of ULBs generally have weak finances that would not allow them to qualify for loans. The TWAD Board has also therefore "not maintained a revenue surplus since many small urban and rural bodies are unable to pay their bills" (Gopakumar 2012 62). The result is a further weakening of TWAD's institutional and financial standing in ways that further disadvantage rural communities in the state.[31] The kind of restructuring that is associated with dominant global norms thus has a more significant impact on small towns and rural communities both by introducing new corporate models of governance and by intensifying the financial marginalization of smaller urban local bodies. Such processes have a stratified effect on state institutions that reinforce the inequalities between the metropolitan city and wealthier urban local bodies on the one hand and rural and small-town communities on the other (Kundu 2001).

If new models of financing have been a key feature of reforms for small towns, the need for the participatory management of water resources has become a dominant discursive frame that has been promoted by global institutions and NGOs and incorporated within national and local state policy approaches to rural India. The primary institutional reforms that have been implemented by the state have been modeled around prevailing global models of decentralization and the creation of Water Users Associations in rural areas.[32] However, Tamil Nadu provides vivid examples of the ways in which programs of decentralization can in effect reinforce or produce new forms of centralized state authority. Consider some of the critical insights of Satyajit Singh (2007), who led the World Bank's Water and Sanitation Program's Rural Team from 1999 to 2002. Singh presents a nuanced critical assessment of attempts at the decentralization of water governance across various states in India. Writing about the case of decentralization in rural Tamil Nadu, Singh documents the ways in which key positions in newly established water committees were staffed by the major rural state bureaucrats of the district and the Tamil Nadu Water Supply and Drainage Board (TWAD). In this framework, while the Village Water and Sanitation Committee is given the responsibility for water governance (ensuring the operation, management, and sustainability of water supplies), the authority over funding, design, and implementation of water infrastructure rests with the conventional structure of the state's water bureaucracy. The result, as Singh notes, is that there is a "system of unclear accountability" in which "the new deconcentrated system uses PRI [Panchayati Raj Institutions] as line agencies of the state as and when it is useful to the state" (2007, 206). This process parallels the ways in which the PWD has retained its authority over Water User Associations in Tamil Nadu (see chapter 2). Decentralization in this context devolves state responsibility to new organizations of local governance while retaining the centralized authority of long-standing bureaucratic organizations.

This reworking of state power is not unique to the case of Tamil Nadu. Rather, it is built into the institutional process of reforms. For instance, the decentralization of rural water governance contains within it an internal contradiction. For example, the Public Health and Engineering Departments have been asked to design their own reforms, and as Singh notes, "It is indeed naïve of the central government to expect the PHEDs to write themselves out of existence! The structure of the implementation of the reforms ensures the sabotage of the reform process itself so there would be a policy reversal" (2007, 199). This sabotage of decentralization points to a need for a deeper rethinking of the question of regulatory reforms in the water sector. The transformation of decentralization into new networks of state power points to the ways in which such institutional reforms contain within them the nodes of centralized power. From such a perspective, the recentralization of state authority of water resources is not simply a form of bureaucratic sabotage but an intrinsic dimension of both the national and global model of reforms that has recentered the authority of state governments. Centralized state control is, in effect, reconstituted at a different spatial scale.

Intersecting Inequalities and the Stratified Space of the "Local"

Given the transformation of institutional reforms into processes that reproduce various forms of state-led extraction and control, such reforms then inevitably become entangled in long-standing socioeconomic inequalities. The state-led process of regulatory extraction becomes enmeshed in the varied forms of socioeconomic stratification that produce enduring structures of inequality both between and within urban and rural spatial locations and communities. While there are new systemic forms of extraction that produce a structured relationship of inequality been the city and the remainder of the state, rural and urban communities are, of course, not homogeneous categories. Poorer communities within cities do not have the same access to water resources as middle- or upper-class communities (Anand 2017; Dasgupta 2015). Such inequalities are reworked in the processes of regulatory reform that are enacted.

For example, a central dimension of Tamil Nadu's Water Resources Consolidation Project was the creation of regulatory mechanisms to manage land acquisition for water infrastructure construction and management. The World Bank incorporated a focus on planned land acquisition and economic rehabilitation as a key component of its funding for the reforms of Tamil Nadu's water sector (spending a total amount of \$5.3 million).[33] While on one hand, the objective of ensuring systematic compensation for individuals and families displaced by infrastructure projects provides an important mechanism for preserving socioeconomic rights, such regulatory reforms have also institutionalized the state's right to displace individuals in the service of developmental goals. A new governmental organization, the Land Acquisition and Economic Rehabilitation Office, was instituted as part of the reforms along with new governmental policies for land valuation by "negotiated settlement" in order to provide "speed and flexibility in determining compensation levels based on full market value and transaction costs for purchase of fully equivalent agricultural land" (WB 1995a, 8). Aspects of the new regulations attempted to address deeper forms of socioeconomic inequality, for instance by including landless laborers within the formal definition of individuals affected by development projects, thus making them eligible for compensation (127). However, the World Bank itself provided hints of limits to its rehabilitation objectives even within the terms of irrigation projects that it funded as part of the WRCP. The project completion report stated that "project-affected persons (PAPs) are as well off or better than their previous situation" but also noted delays in the transfer of lands (Rajagopal 2005).

Another appraisal of the rehabilitation project, while praising the LAER (Land Acquisition and Rehabilitation Office) as an innovative measure, noted, "The separate component for land acquisition and economic rehabilitation worked well for acquiring land, but faced some limitations in rebuilding the livelihoods of those adversely affected by the project" (OEDWB 2005, 2). Consider, further, the details of the process of displacement and rehabilitation through one scheme that was part of the Bank-funded project in Tamil Nadu, the Mordhana Reservoir scheme. As part of financing regulations, a detailed report of the government's rehabilitation plan was submitted to the Bank (ORG 1994). The project, one of nine schemes that were funded, was centered on the construction of a dam and water storage facility that was intended to stabilize irrigation for the area both for irrigation supply and for flood control. The area affected was estimated at 133.62 ha. of land in two villages, of which 46.28 ha. was under private ownership (ORG 1994, 2; the rest was already government land and was under the control of the PWD). A detailed survey conducted as part of the report indicated that the majority of land losers were from low-caste (Other Backward Classes and Scheduled Castes/Dalits) small and marginal farmers (7). The survey provides an important picture of the stratified socioeconomic effects of such small rural infrastructural projects, which are generally rendered invisible in the context of the more visible developmental activities in cities and urbanized areas. At one level, the institutionalization of market-based compensation for land and housing as well as the creation of formal channels for grievances in the process represent a positive regulatory advance in contrast to arbitrary rehabilitation or uncompensated displacement. However, the long-term effects of the compensation are structured in significant ways by the intersections of caste, class, and gender inequality inherent in landownership and therefore in the corresponding implications of rehabilitation. According to the survey, 44 percent of the displaced people intended to spend their compensation on the purchase of agricultural land, and a quarter intended to invest in land development and the purchase of livestock. While these segments of the affected villages could potentially acquire a sustainable livelihood and in some instances benefit from the compensation, the remaining third of the affected population needed to use their compensation for immediate subsistence needs or to pay off debts (10). The process of rehabilitation does not provide any assessment or avenue for the future sustained livelihood for this marginalized section of displaced people, primarily from OBC/SC castes.

The state's management of displacement and rehabilitation was also structured in significant ways by gender. Rehabilitation was structured around gendered definitions of landownership, despite the fact that women play a significant role in both farming and the management of water resources. In the case of the microdynamics of the Mordhana scheme, the survey revealed that eight women expressed a negative impact on their economic standing. Five women indicated that they had to take on wage employment because of the loss of land, and the rest indicated they had to commute longer distances to other villages in search of work (ORG 1994, 15). In addition, the policy of rehabilitation excluded female adult members of affected households. As the report noted, "Major daughters have not been included for MA [maintenance allowance] and RA [rehabilitation assistance], since [a] majority of them get married within 20 years of age and inclusion of them entails [a] lot of complications which would be difficult to tackle for a smaller LA & ER [land acquisition and economic rehabilitation] cell" (38). The regulatory policy reform thus institutionalized a gendered conception of both labor and family that erased the labor of female members of households as well as the fact that long-standing historical patterns have shown the persistence of (often undercounted) female-headed households in rural contexts in India (Agarwal 1994).

Such infrastructural projects intensify long-standing intersecting inequalities of caste, gender, and class despite the best efforts of such reforms to provide for ameliorative measures for marginalized socioeconomic communities. The regulatory mechanisms of the state, of course, always contain the strong and self-evident risk of reproducing the inequalities and exclusions that shape local communities in both urban and rural areas. Institutional reforms that have attempted to produce greater farmer participation in the management of water resources have also tended to reproduce or intensify such inequalities. One in-depth study on the Lower Bhavani Project commissioned as part of the state's assessment of its institutional reforms revealed that village hierarchies and gendered social norms (such as domestic responsibilities and patriarchal resistance to women's participation) posed considerable constraints on the participation of marginal farmers (CWR 2003). The report's survey found that a "majority of Scheduled Caste farmers felt that agency officials discriminate against lower caste men" and that a "majority of women farmers say there is discrimination by officials and felt that [in] the WUA activities males are favored" (CWR 2003, 92). Such forms of social discrimination were compounded by the intersection with class inequalities. For instance, marginal farmers were prohibited from participating by the lack of resources to forgo wages or invest money for travel expenses to attend meetings. Given that the report estimated that in the state of Tamil Nadu as a whole, 73 percent of landholdings were owned by marginal farmers and that marginal farmers were a rising trend (CWR 2003), the obstacles to their participation represent a significant limitation on the WUAs' representativeness.

These patterns of exclusion illustrate the ways in which intersecting inequalities are embedded within decentralized institutions that have been established for the management of water resources in rural areas. Institutional reforms in agricultural areas are enmeshed in long-standing socioeconomic hierarchies in ways that do not expand inclusion or access in the management of water resources. Such hierarchies have produced episodic forms of local protest. In one instance, local villagers from Velliyur attempted to stop Metrowater from purchasing water from their village through both direct social action and legal action (Janakarajan et al. 2007, 56) These protests are often spearheaded by women, as they are responsible for managing household water needs and resources, and in two instances women's organizations were able to successfully stop the sale of water to Metrowater (Janakarajan 2004, 10).

Consider further how the rural-urban relationship that undergirds groundwater markets is shaped by a multilayered set of inequalities of class, caste, and gender. The ownership of land and the natural constraint of whether groundwater is present are critical factors that shape whether farmers are able to benefit from the groundwater markets. Scholarship on rural markets in Tamil Nadu has shown that the sale of water intensifies various forms of inequalities within rural and peri-urban areas. Larger landowners have benefited from the rise in groundwater markets, while landless agricultural laborers who lose employment when land is diverted from agriculture to water extraction are the most adversely affected (Ruet, Gambiez, and Latour 2007).

The growth of water markets has increased competitive water extraction and also exacerbates inequalities between water sellers and water purchasers (Moench, Caspari, and Dixit 1999). Less well-off farmers also accumulate debt when they take out loans for water extraction infrastructure only to find that their groundwater levels are insufficient or depleted by competitive extraction to provide profits (Janakarajan et al. 2007, 58). Gender- and caste-based inequalities that structure landownership have also been reproduced within expanding groundwater markets. The establishment of water markets produces deeper land transformations by transforming property rights in significant ways (Ruet, Gambiez, and Latour 2007). Consider the effects of one Metrowater agreement with farmers in a peri-urban area. Prior to the agreement, while farmers engaged in the private exploitation of groundwater, the water remained in customary terms a common resource, which marginal users such as dependent and semidependent farmers in the area had access to (Ruet, Gambiez, and Latour 2007, 118). Yet, after the agreement, transfers of water within the area were stopped, resulting in "a de facto privatisation of the *access* to the resource, that is, a quasi-privatisation of the resource. The implementation of the agreement pushes towards de-alignment of the property rights structure from something close to common property towards a system that is nearly constitutive of a private regime" (Ruet, Gambiez, and Latour 2007, 119).

There is, in effect, a paradox in this reconfiguration of public goods and private rights. The maintenance of the public supply of water for the metropolitan city area deepens the commodification of water in ways that narrow the public domain of this peri-urban area. Such processes have contributed in significant ways to India's deepening agrarian crisis in the postliberalization era.[34] As one study has shown, in the case of Tamil Nadu, "water marketing villages are experiencing a decline in agriculture from 20 to 95% during 1990–2007, drinking water scarcity (quality wise as well as quantity wise), depletion of the water table from 0 to 6 m bgl during 1971–2007, the necessity of . . . depending on private water, and the related economical burden due to the informal nature of extraction" (Packialakshmi, Ambujam, and Nelliyat 2011, 436). The state's conception of the public good is in the process stratified by a city-periphery model that has become fully entrenched in the postliberalization period.

While the extraction of water is shaped by accentuated structural inequalities between rural and urban communities, urban communities are also of course marked by internal inequalities. High-income groups use bottled water and private water supplies, middle-income groups use hand pumps, and low-income groups use Metrowater hand pumps located on streets (Saraladevi 2013, 152). This class-based differentiation in water infrastructure (see figure 4.3) is also gendered, as women are responsible for the labor entailed in obtaining water for household needs. Socioeconomic status is also shaped by the calculus of electoral politics. Thus, marginalized communities that are politically organized may also use protests to pressure local

FIGURE 4.3. Tamil Nadu Housing Development Water Source, showing the communal water pump from a low-income housing colony under the administration of the Tamil Nadu Slum Clearance Board

state officials. Given that wealthy and upper-middle-class families have a steadier supply of private water resources, it is also the case that lower-income communities that rely more fully on Metrowater may be more likely to protest disruptions in supply. In this context, there are ways in which even poorer communities are stratified in complex ways. For instance, sections of the urban poor that have received state-supported housing through

Tamil Nadu's Slum Clearance Board may have more political leverage than poor communities living in informal settlements.

Consider one stark example of the stratification of water markets and the urban poor in Chennai. In a housing development of the Tamil Nadu Slum Clearance Board (TNSCB) on the outskirts of the city, the sewer and water lines had been breached, resulting in residents receiving contaminated water for a period of two months. One of the women indicated that she had noticed the water had taken on a greenish appearance and she knew something was wrong. She stopped using the water and began buying water in tin cans. However, she said that she continued to collect water from the community pipe (gendered norms mean that women are responsible for ensuring that water is collected for household needs). When I asked why she would still collect contaminated water, she responded that she was collecting it and selling it (interview and site visit, August 18, 2016). This example is a stark illustration of the entangled contradictions of water bureaucracy, water markets, and inequality in Chennai. The delay in the repair of the breached pipes meant that inadequacies in the water bureaucracy compelled this woman to rely on informal private water markets. Yet her ability to sell polluted water through informal private water markets also underlines the deep stratification of poverty. The combination of socioeconomic marginalization and water scarcity produces stratified water markets among the urban poor. Meanwhile, underlying this story of markets, poverty, and survival is a deeper story of institutional cleavages in the water bureaucracy. The engineer in charge of the complex knew about the breach but said he was helpless since he was employed by the TNSCB and the infrastructure was maintained by Metrowater. According to the engineer at the housing site, as an employee of the TNSCB, he was responsible for water infrastructure maintenance within the buildings and homes (interview, August 18, 2016). Thus, the institutional division of authority meant that the on-site engineer could do little to jump-start repairs of the pipes. The institutional cleavages within the water bureaucracy themselves reflect the relationships of power that shape access to water within and between urban and rural communities in Tamil Nadu.

Institutional reforms produce a redistribution of centralized institutional power rather than a shift from centralized to decentralized state governance.

The implications of such shifts in governance have far-reaching consequences for the political economy of the state. Urban governance in this context cannot be understood through a reified lens of a territorially bound city. New modes of urban governance intersect with the historical weight of both bureaucratic and political-economic structures in ways that reconfigure the use of land and water across the divides of rural and urban spaces in the state. The changes produced by economic liberalization do not unfold either according to dominant models of "neoliberalism" or through models of reform that assume a linear reworking of the relationship between the state and private capital. Rather, the politics of water are shaped by the socioeconomic inequalities and institutional relationships of power that stem from the models of city-centered urban development that are being produced by both state practices and private capital investment in the postliberalization period.

This focus on the restructuring of Chennai's water bureaucracy allows us to gain a deeper understanding of the workings of the postliberalization state that are not adequately captured by exceptionalist narratives of bureaucratic corruption and state failure in India. The overdetermined processes that constrain the ability of water bureaucrats to effectively manage water resources are deepened as competition over water resources is intensified by accelerating and unplanned models of urban development in the postliberalization period. Monolithic stories of state failure—whether they are told in terms of incapacity, corruption, or the capture of the government by private interests—are accurate but not sufficient for an adequate understanding of the state.

An adequate understanding of the state requires a deeper understanding of the nature of bureaucratic agency. Take, for example, the case of Chennai's struggles with the management of droughts and floods. Such crises are not new to the city, which has had a long history of coping with floods and droughts.[35] Yet shifting weather patterns and the potential impact of intensified swings between drought and floods that are associated with climate change produce new and daunting challenges for state employees in the city and the state. Consider, for instance the technical challenges of operating the city reservoirs. From a civil engineering perspective, reservoir operation has become increasingly difficult, as there are conflicting objectives of keeping a maximum amount of water in storage to cope with water demands on the one hand and ensuring enough empty storage space for storing flood waters on the other (Anbarasan 2010, 4). Since the state cannot construct new reservoirs for the city, bureaucrats struggle with planning for floods in the winter and drought in the summer. State employees in the water bureaucracy are often faced with managing crises that are the product of state policies that they have little control over. It is this question of bureaucratic agency that I turn to in the next chapter.

CHAPTER 5

State, Class, and the Agency of Bureaucrats

BUREAUCRACIES, SUCH AS THE WATER BUREAUCRACY, ARE OFTEN viewed as nameless, faceless abstractions. In practice, the water bureaucracy is made up of individual state employees who are often left with the task of managing the stresses of water-related crises in the state. Consider the experience of the engineer in charge of Chembarambakkam Reservoir, which he described as he reflected back on his experience during the historic flood in 2015 (interview, January 19, 2021). He recalled being unable to leave the reservoir for a period of ten days while he monitored the flood levels, fearing that the entire reservoir would be breached. He recounted the political pressure of higher-level state employees phoning him continually as they faced mounting public anger from city residents. His own home was completely flooded, and his family had to evacuate to a hotel, but he could not leave his station for a moment because he feared being blamed for the flood damage. As a technical worker without a college degree in engineering, he had remained at the level of assistant engineer and was now only a few months away from retirement. The hidden work of such public sector workers and bureaucrats, who quietly persist in trying to do their jobs both in times of crisis and in routine contexts, usually does not merit much attention in the grand narratives on bureaucratic inefficiency and corruption on the one hand and institutional reform and good governance on the other. Yet "the state" is as much made up of the microdynamics of such practices and employee action or inaction as it is of the grander processes of reform and policy.

Although debates on institutional reforms and questions of governance often focus on narratives of corrupt, inefficient, and recalcitrant bureaucrats in India, a more variegated institutional field is at play—one marked by differences between and within bureaucratic organizations. These patterns illustrate both the political and economic constraints on bureaucratic agency as well as spaces in which bureaucrats have tried to engage in constructive and effective governance. A complete account of the challenges of water governance and the dynamics of institutional reform requires a framework that incorporates a more complex understanding of bureaucrats. Such an account unsettles explanations of failures of governance and regulatory reforms in terms of exceptionalist arguments that locate such failures in a presumed monolithic inertia of the "Indian bureaucracy." The complexities of the bureaucracy in the postliberalization period also lie in the fact that the bureaucracy is itself a target and site of restructuring as well as an agent in the implementation of reforms.

Consider one of the singular historical figures of Tamil Nadu's water bureaucracy, Professor A. Mohanakrishnan. His professional diary contains the following comment on an Advocates' Conference he attended as Tamil Nadu's chairman of the Cauvery Technical Cell: "Very little done. I felt sad on waste of time and money and to satisfy my conscience, have not been joining the costly lunch provided by the hotel and carried my own buttermilk rice these days" (2016b, 72). Mohanakrishnan's observations about a meeting that he had attended to discuss the Cauvery interstate dispute provide a rare glimpse of the personal reflections of a key actor in Tamil Nadu's water bureaucracy. His reflections point to a realm that is rarely addressed in either public narratives or academic writings on India's bureaucratic institutions. Mohanakrishnan refers to his response to the waste of time and money not in the terms of conventional rational actor models of bureaucratic agency or in the traditional terms of social and political narratives of resistance. Rather, he refers to his resistance to the waste of time and money—a familiar microexample for analysts and critics of bureaucratic practices—as a way to satisfy his *conscience*. This response unsettles the customary economic, political, or institutional frames that form the grammar of both public and academic analyses of India's bureaucrats.

The idea of a bureaucrat's conscience has had little discursive space, given the weight of condemnation of the figure of the Indian bureaucrat. Bureaucratic agency in both scholarly works and public discourses in India has largely been described through the control of information, access, and resources that is executed through vast mazes of paper, patronage, and petty power (Gupta 2012; Tarlo 2003). Scholarship in comparative contexts has rightly pointed to the ways in which bureaucratic structures are both embedded within and generative of exclusion, inequality, and what Michael Herzfeld has termed "social indifference" (Herzfeld 1992; Lipsky 2010). The bureaucrat has, often justifiably, come to embody the worst excesses of the state. Yet while the figure of the bureaucrat looms large in such analyses, there is little discussion of the bureaucrat as a complex subject of history. In the Indian context, the subaltern turn of analysis that has sought to present the subjectivity and agency of various social groups in nuanced ways has not extended to the bureaucrat. The figure of the "bureaucrat" in fact encompasses a widely stratified set of individuals and social groups. The top tiers of the bureaucracy range from the elite levels of the state, such as the national IAS cadres, to the top tiers of the state government. However, the full fabric of the bureaucracy consists of a broad range of intermediary employees of varying privilege and status, from generalist administrators to specialists such as engineers and technical experts to lower-tier public sector employees.

In many ways, the analytical and theoretical gaps in analyses of bureaucrats are an understandable effect of the structural location that the bureaucracy occupies in the political economy of India. The bureaucracy is the arm of the state that executes policies of development and is the institutional field responsible for the (often ineffective) implementation of state policies. The bureaucracy is inextricably wound up in both the historical weight of longstanding developmental failures and the legacies of singular but spectacular crises, such as the Emergency period. Critics of the effects of India's developmental state have aptly shown the ways in which the bureaucracy has encoded various forms of power relations that have transformed poor communities into targets and casualties of state programs that failed to successfully ameliorate their lives (Bardhan 1984; Chatterji 2006). Public discourses on corruption and popular anticorruption movements that have arisen in recent years have tended to focus on the corruption of bureaucrats.[1] Meanwhile, India's economic reforms advocate for a scaling back of the bureaucracy. The agency of bureaucrats is thus entangled in the very real web of political and socioeconomic deficiencies produced by state policies on the one hand and political and public discourses on the bureaucracy on the other.

The conceptual challenge of addressing the complexity of bureaucratic agency requires an approach that addresses the ways in which bureaucrats are enmeshed in such broader political and socioeconomic patterns and are themselves complex social actors. Consider the historical formation of bureaucratic workforces. At the macro level, state investment in an expanding bureaucratic apparatus in the early decades of independence became a significant dimension of middle-class formation in postcolonial India. In contrast to the elite bureaucrats of institutions, such as the IAS, the vast majority of bureaucrats consisted of the middle and lower tiers of the middle classes. These sectors were employed in public sector enterprises or responsible for delivering services at the local level. Public employment has long been a central avenue for middle-class individuals as well as for upwardly mobile members of the lower-middle classes and working classes. An adequate understanding of bureaucratic agency requires an analysis of the distinctive nature of this class-state relationship, where bureaucrats are both the product of and the agents of state policies.

India's bureaucratic field is shaped by three significant dimensions. First, the historical formation of institutional rules and norms—including both long-standing continuities and periods of institutional change in the postindependence period—continues to shape the postliberalization period in significant ways. Second, the bureaucratic field is shaped by the complexities of middle-class formation in contemporary India. An adequate understanding of India's bureaucracy thus rests on a nuanced understanding of the differentiated material and symbolic dimensions of the middle classes and the role of the state in shaping middle-class formation in India. Finally, an adequate understanding of the bureaucracy requires a more nuanced field of the agency of bureaucrats, who must maneuver within the complex institutional, political, and economic structures that shape their employment.

Historical Legacies and the Political Underpinnings of India's "Steel Frame"

The historic image of India's bureaucracy is often captured by Sardar Vallabhai Patel's characterization of the Indian Administrative Service as the steel frame of newly independent India. Cautioning that "you will not have a united India if you do not have a good All-India Service which has independence to speak out its mind,"[2] Patel founded his advocacy of the IAS on images of a highly professionalized and independent institution that embodied the bureaucratic potential of the newly independent and interventionist Indian state. In this context, contemporary representations of a dysfunctional bureaucracy are often presented through a narrative of growing decline of this ideal through the increasing politicization, inefficiency, and corruption that unfolded in later decades. The deterioration of the IAS and the state bureaucracy in general is often marked by the systemic politicization of Indira Gandhi's practices during the Emergency period.[3] Yet while Indira Gandhi's politicization of the bureaucracy during her Emergency Rule is well known, this acute form of politicization that began in the mid-1970s must be contextualized in a broader context, where India's bureaucratic field has been shaped by political processes since the inception of the Indian Civil Service (ICS) in the colonial period.

The ICS was a central instrument for the exercise of colonial state power. The image of the ICS as a "steel frame" that would later come to be popularly associated with the IAS in independent India was in fact a metaphor for the colonial state's political project of retaining power and maintaining law and order. As Lloyd George would note in his speech to the British Parliament in 1922, "If you take that steel frame out of the fabric, it would collapse. There is one institution we will not cripple, there is one institution we will not deprive of its functions or of its privileges; and that is the institution which built up the British Raj—the British Civil Service of India" (quoted in Benbabaali 2008). Scholarship on the Indian Civil Service has provided in-depth historical analyses of the political dynamics, which included the politics of racial stratification between British and Indian employees, the authoritarian nature of ICS rule, and active political interference in the face of a rising nationalist movement (Misra 1977; Nayar 1969; Potter 1996). The colonial nature of the ICS meant that it was oriented toward law and order and revenue collection in order "to limit the role of government, to promote stability by minimizing change, to co-ordinate the activities of government, and to provide a tight chain-command control over governmental actions and personnel" rather than toward developing welfare-related dimensions of the state that would serve broader societal needs (Nayar 1969, 10). This did not inoculate the ICS from processes of politicization, which included transfers of ICS officers due to political interference, the pursuit of partisan objectives, and the use of district administration structures as a means of repressing the Indian nationalist movement (for instance by increasing taxes on villages that supported the Congress organization (Potter 1996). Consider one example, where "in January 1921, the water level of the Godavari River fell rapidly, and it was necessary to reduce the area of second crop irrigation. The Collector, Bracken (ICS, Madras) and Wadsworth [Additional District Magistrate, ICS Madras], when making decisions on this, gave preference to those villages that had not joined the non-co-operation movement, 'thereby demonstrating in the most convincing fashion that it paid to be loyal'" (Potter 1996, 36).

This deep politicization of the ICS, as is well known, produced significant opposition to retaining the institution after independence. The point at hand is not that contemporary forms of corruption and politicization of the bureaucracy have been causally produced by colonial practices but that such processes of politicization do not emerge in a historical vacuum in the later decades of the postindependence period. The use of the administrative services both for partisan ends and for the purpose of repressing political dissent has a long historical legacy that has produced deep-seated interconnections between the bureaucratic and political fields in India.

In the early years of independence, the turn from the colonial to a newly independent developmentalist state would require the public bureaucracy to "undertake new tasks and discharge new functions which were till 1947 beyond its scope . . . [as] the bureaucratic apparatus of the security state was being gradually transformed into an instrument of [a] welfare state" (Prasad 1974, 29). In the process, the administrative apparatus became a central vehicle for the creation of relationships of patronage and dependence between the state and various socioeconomic groups. Given the single-party dominance of the Congress in the early decades of independence, political dynamics were built into this emerging role of the bureaucracy (Bhambhri 1971).

These early patterns of politicization of the IAS were deepened and systematized in the 1970s and 1980s. As the single-party dominance of the Congress was unsettled, the civil service increasingly became a site of political contestation, particularly as the rise of regional parties began to increase political tensions between the central government and the states. Such events produced new pressures created by the Emergency, local state politics, and regional nationalism (L. Rudolph and S. Rudolph 1987, 75). Prime Minister

Indira Gandhi's rule, and the authoritarian excesses of the Emergency period, of course, represented the weightiest and most visible period in which the civil service, and the bureaucracy in general, was politicized. Indeed, the use of the bureaucracy to implement some of the most repressive state policies, such as the forced sterilization program, has imprinted the bureaucracy with an indelible mark of repressiveness (Tarlo 2003). However, the formation of a politicized bureaucracy has also unfolded in ways that have more broadly and systematically connected the bureaucratic and political fields in the postindependence period. Such concerns were intensified in the 1970s as "loyalty to the party in power became part of their [IAS officers'] reward structure" and affected promotions, postings, and transfers of civil servants in more systematic ways (Potter 1996, 156). Political patronage in effect became an intrinsic component of the institutional rules of the bureaucracy in ways that surpassed the specificities of the Emergency period or the Congress party's era of centralized control.

As the deployment of political capital became a structural component of the bureaucratic field, the politicization of the bureaucracy cut across both national and local levels and across political party. While the IAS represents an elite and relatively small component of the Indian bureaucracy, this politicization signified a broader incorporation of political capital within the reward structure of the bureaucracy. This structural feature of India's bureaucratic field led author and well-known journalist Prem Shankar Jha to characterize bureaucrats as a service-oriented "intermediate class" because of the systematic ways bureaucrats used their position to in effect extract money for the delivery of services. As he would caustically argue,

> Because they accept money in return for services, and they are members of the intermediate class insofar as the value of the services they render condoning black-marketing, bootlegging, smuggling, colluding in the evasion of excise levies, or speeding up the process of obtaining official sanctions, increases with the intensity of the shortages being experienced by the economy . . . the bulk of the police force, and the majority of the staff of the economic and technical departments of the Central and state governments, who come in direct contact with the public, can be considered members of this class. (1980, 100)

This acerbic depiction captures the ways in which the deployment of political capital became a more generalized form of patronage that was systematically incorporated within the reward structure of the bureaucracy—a process of institutionalization that is overlooked by the more generalized or homogeneous discourses on cultures of corruption.

Reforming the Bureaucracy in Postliberalization India

The question that then arises is whether and how the rhetoric of reform has begun to transform the bureaucracy. While global discourses on good governance have become part of the Indian rhetorical lexicon on reforms, such reforms have in fact focused more on conventional processes of workforce restructuring, which are typical of structural adjustment policies. Developments since the 1990s reflect a restructuring of large sections of the bureaucracy that have paralleled forms of industrial restructuring that have targeted workers through cutbacks in employment without restructuring the underlying political nexus that has produced historical legacies of bureaucratic dysfunction. At both the central and state government levels, workforces have shown a gradual but steady reduction in employment in the first two decades of the postliberalization period. In addition, vacancies have not been filled.[4] Meanwhile, financial pressures on local state budgets have meant that state governments have had to consider finding ways of managing the costs of state government employees. In the case of Tamil Nadu, a state well known for its populist rhetoric and policies, the state government has nevertheless restructured its government workforce and also attempted to make changes in its compensation (for instance by restructuring pension funds for state employees).[5] Tamil Nadu's state government workforce dropped from 587,111 in 2007 to 527,790 in 2017 (Sheelapriya 2008; Anbu 2016). While the scaling back of the state bureaucracy is a central component of the reforms agenda, this retrenchment does not address the deeper issues of the systemic incorporation of political capital within administrative reward structures of the bureaucracy.

Let us return, for instance, to the IAS, the most privileged sector of the bureaucracy. Recent research provides a vivid example of this process of restructuring, which has been skewed toward cutbacks in employee benefits without a change in the politicization of the reward structure. The historical trend of using transfers to exert political pressure on bureaucrats has continued well into the postliberalization period (Iyer and Mani 2012).

According to a Government of India survey, 52 percent of respondents indicated that they believed "that the postings to important posts and sought after stations are not decided on the basis of merit while 58% officers feel that the transfer orders are not issued keeping in mind the specific needs of the concerned" (MP 2010, 55). The survey also indicated that respondents believed that nepotism and political influence played a key role in enabling civil servants to gain access to top-tier positions. While the underlying party-administrative nexus of the bureaucracy has remained unchanged, economic shifts in the postreforms period have affected the status of IAS employment. As private sector white-collar employment has become more lucrative, the prestige and monetary rewards of IAS employment have shown a relative decline, particularly in contrast to upper-tier private sector employment (Vaishnav and Khosla 2016, 10).

The combination of the continued politicized and patronage-based foundation of the bureaucracy with declining prestige and compensation has further entrenched the institutionalization of extralegal monetary compensation as part of the bureaucracy's reward structure. Understanding this process in ways that do not invoke simplistic languages of corruption is not an attempt to rationalize the damaging effects of bribery and graft but a means of grasping the institutionalized, structural foundations of such rent-seeking behavior. As Milan Vaishnav and Saksham Khosla have noted, "Endemic political interference can lead to rent-seeking behavior even for honest officers, who might feel forced to comply with questionable demands from superiors for fear of being punished. Furthermore, uncompetitive public-sector salaries (not to mention years of foregone wages as candidates devote an increasing amount of time to passing the civil services exam) encourage officers to make extra money while in office" (2016, 12). Processes of liberalization have thus left the political underpinnings and corresponding patronage structures of the bureaucracy firmly in place.

The continued political dysfunctions of the bureaucracy have led to two interrelated central lines of scholarly inquiry. On the one hand, scholars have focused on an analysis of the state as a political formation that is embedded in practices of corruption and relations of patronage (Das 2001; Chandra 2015; Gupta 2012, 2017). Political scientist Kanchan Chandra, for instance, has made the important argument that policies of liberalization have simply redirected state power so that "the retreat of patronage from some areas of the economy has been accompanied by a relocation to others"

(2015, 46). While the Tamil Nadu state government has reformed some aspects of the water bureaucracy, as we have seen, state power has been expanding in the management of water resources through the licensing of groundwater wells, the regulation of land acquisition, and the expansion of the power of institutions such as Metrowater.

What remains an understudied area of such shifts in the state in the postliberalization period is an understanding of the relationship between this restructuring of the state and the processes of class formation of the bureaucratic workforce. One of the central dimensions of the Indian state both in the colonial period and in the early decades of independence has been the way in which the state produced and shaped middle-class formation. Historically, state subsidies of higher education and the role of the state as an employer of large sections of the middle classes meant that the state played a central role in the production and support of India's middle classes. However, in recent years, the main focus of the reforms of India's bureaucracy has been on the gradual but steady reduction of government employees. An adequate understanding of the restructuring of the bureaucracy thus requires a closer analysis of the bureaucratic field as a realm that represents this changing class-state relationship.

The Indian State and Bureaucratic Middle-Class Formation

One of the most astute understandings of the relationship between the state and middle class in recent years emerged not from scholarly analyses of contemporary India but from Modi's 2014 electoral campaign and his deployment of the idea of a "neo-middle class."[6] In contrast to both celebratory public marketing presentations of an expanding, successful postliberalization middle class and academic scholarship that has reinforced conceptions of this middle class as intrinsically linked to (if not a product of) market-led growth and consumption, Modi's rhetoric captured both the limits of access to middle-class status and the continued significance of state support for large sections of the middle classes in India. As the BJP's party platform noted,

> India has a large middle class with immense understanding, talent and purchasing power. In addition, a whole new class has emerged. Those who have risen from the category of poor and are yet to stabilize in the middle class,

the "neo middle class." This class needs proactive handholding. Having moved out of poverty, their aspirations have increased. They want amenities and services of a certain standard. They thus now feel that Government facilities and services are not up to the mark, and hence resort to the private sector for things like education, health and transport. This is obviously costly, putting the neo middle class into a daily dilemma. As more and more people move into this category, their expectations for better public services have to be met. We have to strengthen the Public Sector for providing efficient services to our citizens (BJP 2014, 17).

At one level, Modi's campaign rhetoric, in effect, pointed both to the symbolic power of the promise of upward mobility that has been embodied in images of India's "new" postliberalization middle classes and to the limits of this promise of access, as large segments of society (including sections of the middle classes) have not benefited from wealth generated by new economy jobs within the services and informational technology sectors (Fernandes 2006). At another level, his rhetoric underlined the role of both the state and the public sector as a key foundation of support for these segments of the middle classes. Modi's rhetoric effectively cast new policies of economic reform through historical state languages of development in ways that both grasped and capitalized on the ways in which large segments of the middle classes continue to rely on and to demand various forms of state support—a reliance that rests on the historical role of the state in shaping middle-class formation in India.

A focus on the bureaucracy as a site of middle-class employment provides the analytical space that can grasp the contradictory socioeconomic location of bureaucrats and their relationship to the state. The contradictions inherent in this location are shaped by the ways in which bureaucrats are entangled in a relationship with the state that is both an extractive relationship that allows for the appropriation of resources (Bardhan 1984) and a subordinate relationship of dependence through the conditions of employment. This stratified relationship complicates critical and analytical discussions of the state as an *employer*—whether in the bureaucracy or more generally in public sector enterprises (Ganguly-Scrase and Scrase 2009).

Too often, criticisms of the public sector and the bureaucracy are implicitly or explicitly tied to normative views that advocate modes of material public disinvestment through specific models of economic reform liberalization rather than much-needed correctives to the organization, utilization, and distribution of public resources. Indeed, it has long been easy to deploy images of a bloated and inefficient public sector and a corrupt bureaucracy, given the kinds of relationships of extraction that have haunted the state's management of economic resources. However, the restructuring of the state in the postliberalization period also entails a restructuring of a sizable workforce with implications both for the employees and for the nature of class inequality. Public sector employment in both governmental and industrial occupations has steadily declined since the 1990s. Shifts in the postliberalization period have lessened the significance of public sector employment both in terms of size and as a marker of status, as top-tier private sector white collar jobs have become increasingly lucrative while top-tier jobs in the IAS have not kept pace. However, the impact of this downsizing of public sector employment must be understood largely in terms of its effects on the lower socioeconomic strata of public sector employees (Nagaraj 2014). Given that the nonelite middle classes are often more dependent on state employment than the upper tiers of the middle classes, who have been able to transition to more lucrative new economy private sector jobs in the postliberalization period, the restructuring of the public sector has reduced the security of less-privileged middle-class individuals without unsettling the primacy of the dominant proprietary professional middle classes. Meanwhile, as R. Nagaraj has argued, while large landowners and regional elites have engaged in a "pragmatic use of the public sector [that] seems to be almost entirely driven by electoral calculations," the broader segment of middle classes has resorted to individualized strategies using patronage networks with a "hope to secure individualized gains from a plethora of sub-optimal government welfare programmes, however meager they might be" (2015, 45). The primary effect of cutbacks in the public sector has been to reduce the socioeconomic security of less-privileged sectors of the middle classes rather than to correct the extractive dominance of governmental elites.

Consider the case of state employment in Tamil Nadu. While Tamil Nadu has embraced many dimensions of liberalization and has actively sought private investment, the state government has also deviated from the ideological norms of liberalization through a heavy reliance on populist politics by both major political parties in the state, the DMK and the AIADMK.

Former chief minister Jayalalitha's regime, in particular, drew on gendered constructions of motherhood that consolidated her own symbolic-political representation as mother or "Amma," as she was popularly known, to systematically deploy a range of maternalistic populist programs, subsidies, and entitlements. Such policies ranged from the establishment of "Amma weekly markets" to "Amma Canteens" to the distribution of "Amma water" to the provision of bus passes to college students and the establishment of health insurance schemes.[7] However, in contrast to such visible and substantive welfare provisions, the state government nevertheless engaged in steady cutbacks of its public sector workforce, in line with broader national trends of the restructuring of the state bureaucracy and public sector.

The centrality of governmental employment for the middle classes is underlined by that fact that despite such cutbacks, public sector employment remains a desired avenue for the middle classes in the state. As one report put it, in 2017,

> Greater Chennai has recorded the highest number of registrations as there are 5 employment offices across the city. "More than 9 lakh registrations were made in the last financial year in Chennai city alone as the population is high compared to other districts. Even within Chennai, it is the Professional and Executive Employment Office (PEEO) that has seen the highest number of registrations, 4.96 lakh, till March 31, said an employment exchange official; 12.26% of the total registrations in the state happen in Chennai-based employment exchange offices. Experts say registration of graduates and engineers is an indication that they are not able to get employment in the private sector.[8]

By February 2021, the number of employees from the educated middle classes (consisting of graduates and postgraduates) looking for government employment through the employment exchanges in Tamil Nadu was estimated at 2,616,098 (EES 2021). While unemployment is a key factor for the high number of educated middle classes on the employment exchange rolls, one employee of the exchanges noted that public sector employment is the first choice of job seekers coming to the exchanges, despite a paucity of jobs in this sector (employment exchanges also direct candidates to private sector vacancies) (interview, January 9, 2017). According to the Tamil Nadu Labour Department's 2016–2017 Report, in the preceding five years, 77,696 job seekers were placed in government jobs through the exchanges, while 142,144 were placed in the private sector (Kafeel 2016, 80), confirming the department's observation that the opportunity for employment "in the public sector is dwindling and private sector is increasing due to rapid industrialization in the State" (83). Such trends show that the restructuring of Tamil Nadu's public workforce has taken place in accordance with policies of economic reform in ways that are hidden by the public populist strategies of successive state governments in the postliberalization period. Such temporary populist welfare provisions mediate but do not displace structural shifts in the nature of the postliberalization state.

Broad shifts in the employment of middle-class bureaucrats point to the need for a closer analysis of the bureaucracy as a workforce rather than as a mere arm of the state. This restructuration of the bureaucratic workforce raises the question of the agency of bureaucrats. The interplay between structure and agency plays out through questions of employment, social agency, and subjectivity in the water bureaucracy of Tamil Nadu's Public Works Department. Such a perspective allows us to thicken our understanding of bureaucrats as political and historical subjects rather than as the objectified instruments of state power.

Rethinking the Figure of "the Bureaucrat"

The Public Works Department in many ways encapsulates the complexities and contradictions of India's postliberalization state. The bureaucratic institution, one of the oldest in the country, embodies both the historical legacies of the colonial and postcolonial state and more recent processes of restructuring that have unfolded in the postreforms period. This is materially encapsulated in the physical infrastructure of the department. The head office of the Public Works Department in Tamil Nadu is housed in the historic colonial building. The once imperious building is relatively empty, in contrast to images of a bloated bureaucracy that still permeate contemporary discourses on the state. Meanwhile, the Institute for Water Studies, created through reforms associated with World Bank financial support, shows some signs of the high-tech turn to the mapping and management of water resources through remote sensing and GIS (geographic information system) technologies. While the institute is authorized to hold a total staff strength of seventy-eight employees, this more modern building is also relatively sparsely populated.

Middle-class employees in the water sector of the Public Works Department (all subsequent references to the PWD address the Water Resources Organisation of the institution) consist primarily of technical workers and engineers. However, in contrast to engineers and employees of Metrowater and TWAD, these employees are not directly involved in the provision of water resources to consumers in the state. The PWD represents a segment of the bureaucracy that is not primarily publicly embroiled in the conflicts, inequalities, and relationships of power that are associated with the "street-level bureaucracy" (Lipsky 2010) in Chennai. Existing studies of water politics and the bureaucracy in India have tended to focus on this consumer-related dimension of the bureaucracy within metropolitan areas and have illustrated the ways in which socioeconomic inequalities shape this dimension of the municipal bureaucracy's treatment of urban citizens.[9] However, a sole focus on the service-level bureaucracy within metropolitan areas also produces generalizations about the state, corruption, and inequality that are based on one aspect of a more complex bureaucratic institutional field. An exclusive focus on the service bureaucracy limits the analytical space needed for both a broader understanding of governance and an in-depth understanding of both the structural and agentic dimensions of bureaucrats.

Consider the major thrust of the PWD's bureaucratic charge—the production and deployment of technical expertise both in field settings and at the level of planning. Engineers and technical workers are primarily designated either with the responsibility of managing and operating the physical infrastructure of water sources in Tamil Nadu or with the task of overseeing and developing technical schemes designed to manage the state's water supply. Meanwhile, the public interactions of the PWD are structured by urban-rural cleavages. While the irrigation wing of the PWD is directly charged with providing water resources for farmers, the department is not tasked with interacting with either urban or rural household water consumption. Beyond this, the PWD's interaction with members of the public is linked to questions of land acquisition and the need to protect public land and infrastructure from encroachments. Patterns of rent-seeking behavior have historically been concentrated on more subtle practices in the context of the physical construction of water-related infrastructure projects. As one government report noted, the "scope for corruption" within the PWD rested with a series of project-based practices such as "quality enforcement, paying contractors bills on time, [the] award of works and [the] negotiation of rates" (ARC 1973, 67).

Bureaucratic corruption in Tamil Nadu has been in line with broader national patterns. Public anger and charges over corruption that are endemic within India's water bureaucracy are oriented toward the various components, with particular intensity directed at the utility Metrowater, the Chennai Metropolitan Corporation, and the PWD. In a judgment indicting the Chennai Corporation on widespread corruption, the Madras High Court, for instance, directed the corporation to "file a report explaining the nexus, collusion and corrupt activities largely found between the officials of corporation, police, local politicians, electricity board authorities and metro board authorities."[10] Public complaints about the slow work of the PWD on the maintenance of infrastructure (for instance through desilting) abound in media reports.[11] Or, to take another example of such complaints, in the context of the state's drought, where Chennai residents were relying on water from Metrowater tankers, one individual claimed,

> There is an unspoken corruption in the recently introduced online system for water bookings by the Chennai Metro Water. When I sent my son to check the status of water supply in tankers to Santhome Water Tank Shed (near Cemetery), he was informed that it would take ten days, after booking on line, as there was a huge backlog of bookings. The officials had then demanded Rs. 1,500 for a 9000 litre tank, rather than the usual amount of Rs 600 in order to get the tanker early. The incident explains the prevailing corruption and official inaction in the department. The officials had, in their defense, said that their service is better than the Rs. 2,000 that would be paid to the private tanker.[12]

Public discourses in Chennai are in line with national trends where rumors and complaints about everyday corruption abound and are in effect a component of the making of urban middle-class identity. However, discourses of public corruption that are a continued part of the systemic rent-seeking practices of bureaucratic institutions are only a limited part of the broader dynamics of bureaucratic workforces in the water sector.

Bureaucratic employment in the PWD is a complex field that is marked both by the long-standing reproduction of social stratification and by emerging spaces of social change that are often hidden in stereotypical views of the bureaucracy. As is conventional in bureaucratic organizations, employment is structured through highly stratified internal hierarchies, with the rank of chief engineer representing the pinnacle of achievement within the organization. A vivid symbolic visual representation of the power of this position is embodied in a board in the chief engineer's office at the Institute for Water Studies that lists the names of all the individuals who have occupied this position since the inception of the institute in 1993 in the "Public Works Department Roll of Honor." At the higher levels of the bureaucratic structure, engineers speak of this sense of prestige through a historical lens that is explicitly linked to the significance of the department in colonial times. Senior engineers that I interviewed repeatedly emphasized the historic nature of the department, pointing with pride to both the colonial and postcolonial history of the department as emblematic of a long legacy of engineering achievements in the service of national development. For employees, the creation and management of "public works" in this context were not simply a product of dispassionate technocratic expertise but an embodiment of status and honor.[13]

The broader workforce of the organization reproduces the internal hierarchies and forms of stratification that are typical of workplace settings. In the early decades of independence, the PWD's workforce rapidly expanded in ways that mirrored national trends in the expansion of the public sector. By 1973, the organization had a staff of 2,861 employees (ARC 1973, 16; this consisted of 3 chief engineers, 19 superintending engineers, 124 executive engineers, 665 assistant engineers, and 2,050 section officers). Despite the expansion of the workforce, rigid forms of stratification within the bureaucracy produced various sources of employee workplace dissatisfaction. Engineers, for the most part, had limited space for upward mobility within the organization, and promotions to the higher position of superintending officer (and, in fewer cases, of chief engineer) would generally occur for individuals when they were close to retirement (18). Recent years have seen a restructuring of the workforce again, mirroring national postliberalization trends of the reduction in public sector employment. While, on occasion, employees have unsuccessfully tried to oppose dimensions of this restructuring,[14] the reduction of the workforce has primarily occurred through the practice of not filling vacancies or by hiring employees on a temporary (project-contingent) contract rather than as part of the state's permanent workforce. In the postliberalization period, the obstacles for upward mobility in the department are further structured by internal variations of class distinction that rest on educational capital. There is, for example, a significant difference between employees with engineering degrees and those with technical diplomas, with strong limitations on the upward mobility of the latter group. Socioeconomic inequalities within the middle classes that are structured by credentialing practices are reproduced within the department's workforce.

Within the context of such institutional hierarchies, there are also emerging spaces of change within the department. For example, the PWD has employed significant numbers of women, in accordance with the state government's policy of providing reservations for women in governmental jobs.[15] In the PWD's head office, eleven of thirteen employees were women, while at the Institute for Water Studies, thirteen of eighteen engineers were women. One female engineer estimated that 40 percent of all PWD engineers are women.[16] Within this measurable progress in gender equality, there remain various forms of stratification and inequality. For instance, the major posts of chief engineers continue to be held by men. Meanwhile, there are more men employed in field-based positions compared to higher levels of women employed within the head office and IWS in Chennai.

Women engineers whom I interviewed presented a complex understanding of their employment. For instance, one female assistant executive engineer questioned conventional ideas of choice and said that her employment "was more about her parent's choice" rather than her own. Another female engineer viewed this in generational terms, noting that younger women had more space to make their own career choices. Their narratives also in many ways reflected dominant middle-class constructions of respectable employment and status. As one woman employed as an engineer noted, their employment choices were mainly structured by their performance on central exams. As she put it, "Medical is first choice. Engineering is second choice. Those are the two fields you can get a good job" (interviews, January 19, 2017). Nevertheless, the PWD's employment patterns reflect an example of the success of the state government's reservation policies in changing gendered patterns, particularly given the constraints on women's entry into STEM-related fields in comparative contexts. As one women engineer said in response to a question about gender and engineering, "Gender differences don't matter. Work is work" (interview, January 19, 2017).

Spaces of change are also evident in the ways in which dominant and disciplinary expertise in civil engineering has shown signs of shifting away from the emphasis on the construction of large-scale projects, such as dams and reservoirs. In part due to the scarcity of water sources and of land that can be tapped through such megaprojects, the PWD's focus has been on smaller infrastructural endeavors, such as the construction of check dams (interview with joint chief engineer of irrigation, PWD, January 10, 2017). This is a significant shift, given the ways in which the engineering field has focused on large dams as a central means of harnessing water resources. However, such shifts are not purely instrumental. The PWD, for instance, engages in collaborative work with the Centre for Water Resources at the Tamil Nadu's premier engineering university, Anna University. The focus of this center is on innovative, interdisciplinary approaches to water management that place a significant emphasis on questions of sustainability and social justice. Indeed, internal reports and studies of the PWD often contain important and critical discussions of land management and participatory practices. This is not to imply that such innovative work translates easily into policy shifts. In fact, political obstacles and a fragmented institutional landscape mean that such innovations often rest on the sidelines or face challenges in adequately integrating local participation and knowledge of communities being affected by developmental projects. Local inequalities of caste and class and long-standing interests of local contractors and leaders pose significant hurdles to the effective implementation of such participatory models of water management (Mosse 2003, 283). Change and effectiveness within state institutions is foreclosed through complex configurations of political dynamics, socioeconomic interests, and institutional practices.

Consider, for instance, the ways in which various forms of political intervention constrain the operations of the PWD. In the early decades of independence, scholarship on the bureaucracy (and the IAS in particular) debated at length the tension between generalists and specialists (Nayar 1969). However, this tension was one that existed not simply within the IAS but between IAS cadres and technical experts. As Tamil Nadu's administrative reforms report noted, there was a strong divide between engineers and the PWD wing of the secretariat that "often led to a lack of understanding between the policy-making and programme executing wings of the department" (ARC 1973, 12). The secretariat, in this context, would be a central site where political pressures and considerations would impact decisions and reproduce the state's long-standing role of using infrastructural projects to consolidate electoral support. In recent years, intensified periods of drought and floods only serve to accentuate the politicization of the management of water resources and infrastructure.

Or, to take another example, access to employment in the PWD is handled by the Tamil Nadu Public Service Commission and is based on both performance on central examinations and performance on an in-person interview. However, the composition of the Public Service Commission has itself become politicized. For instance, in 2016, the Madras High Court canceled the appointment of eleven members of the commission, stating that the appointees were members of the ruling AIADMK party who had been appointed without transparency and without following Supreme Court guidelines. As the justices put it, "This appointing process or lack of it was on account of a misconception that the appointment to the post of members of the Public Service Commission was part of the spoils system based on the patronage of the State government."[17] Access to employment in the PWD is shaped by this kind of politicization of external bureaucratic institutions and the relationships of political patronage that mold governmental practices.

Individuals attempting to produce change or to resist the complex political and institutional fields that structure bureaucratic employment must therefore negotiate a precarious environment. Despite the wealth of writing on the bureaucracy in recent years, less attention has been paid to the agency and subjectivity of bureaucrats who persist with such endeavors.[18] Bureaucrats are unlikely subjects who do not fit well within the classic focus on subaltern subjectivity and agency.

Identity, Everyday Practice, and the Possibilities of Ethical Agency within the State

Contemporary work on India's bureaucracy rarely presents an in-depth study of the work experiences or lives of employees. In the colonial period and in the early decades of independence, some autobiographies provided a window into the work experiences and lives of IAS officers.[19] Much of the scholarly writing about India's bureaucracy in the early decades of independence has focused on the elite cadres of the national bureaucracy, such as the IAS. The segment of the middle classes that was tracked into the upper echelons of national government employment drew primarily from existing middle- or upper-middle-class social groups. These segments of the middle classes tended to be from urban metropolitan backgrounds and had access to elite middle- or upper-middle-class schools (see Mangat Rai 1973). However, it is the regional bureaucracies, and less-studied state institutions, such as the Public Works Department, that began to provide avenues for upward mobility for individuals seeking access to middle-class status in postindependence India.

Early in-depth sociological studies of regional bureaucracies reveal a complex set of organizational, cultural, and political factors that produce the kinds of institutional problems that are reduced to a static image of corruption. G. K. Prasad's (1974) study of the governmental secretariat in Bihar based on fieldwork conducted in 1963–65 detailed deep-seated problems regarding the ways in which responsibility was delegated within deeply hierarchical organizational cultures. Writing in particular about the morale of bureaucrats based on field interviews, Prasad noted that local bureaucrats "remarked that while on the one hand merit, though valued very high, did not receive due recognition in government service, on the other, widespread corruption, which acted as a barrier to the efficient working was encouraged in a very subtle way" (98). Such sociological studies have shown a significant degree of discontent of employees in various bureaucratic organizations. Writing in a similar time period, A. Prasad (1976) produced an in-depth study of one of the central bureaucratic figures of the twentieth-century developmental state, the block developmental officer. The study reveals high degrees of dissatisfaction with both the organizational culture and practice and material terms of employment (such as salary and promotions).

Despite sociological evidence of complex problems with organizational and employment conditions, most public and academic discourse focuses primarily on bureaucratic corruption. In recent years, media coverage in India has concentrated on exposing corrupt bureaucrats through often vivid exposés caught on camera or by secret cell phone recordings. Such narratives converge with both proliberalization academic writings that have sought to cut back and reform state bureaucracies and postcolonial theories that have focused on modern state power and the subjugation of citizens. Bureaucrats in this web of narratives rarely emerge as complex individuals whose lives can both illuminate our understanding of state structures and exceed these paradigms by providing a more intricate understanding of the bureaucrat as a subject of history. Professor A. Mohanakrishnan's career at the Public Works Department provides a unique opportunity to gain such a perspective on the otherwise broad sweep of analytical categories such as the "state" and the "bureaucracy."

Prof. Mohanakrishnan's career at the PWD and in government service covered a seventy-year period close to the entire span of postindependence India. He worked in the PWD, gradually rising up from a junior engineer to the highest post of chief engineer from 1947 to 1984. He then served as a governmental adviser and expert in a number of significant positions in the decades since his retirement, finally concluding his career on December 31, 2012. Prof. Mohanakrishnan was also distinctive in his service, as he maintained a strong interest in academic research, writing, and teaching. His career encompassed five years of teaching at Anna University and a sustained period of administrative work at the university, including chairing the department of civil engineering. His receipt of an honorary degree from the university would give him the official title of professor. His academic bent produced a rich set of writings, including technical histories of major infrastructural projects, such as the Mullaperiyar Dam and the Telugu Ganga/Krishna Water Supply Project (whose design and implementation he executed) and a detailed autobiography and professional diary that provides rich insights into the inner workings of bureaucratic life. What emerges from this rich and layered career is a portrait of a bureaucrat that is far from stereotypical images that populate both popular and academic writings.

In his daily recordings of his work experiences, Prof. Mohanakrishnan provides an account of a presentation he made at a seminar on employment opportunities for students at Anna University on July 29, 1983. He describes delivering a presentation at the seminar that opens with the statement, "Government Work is God's Work!" (2016a, 326). The reaction, as he describes it, was that "there was a loud uproar, which I did not mind, and made a convincing speech for 10 minutes. I said what I believed, and I still consider I had done nearly sixty years of unbroken service to the Government before laying down office and through the Government to the State and the people only with God's help" (326).

As with most of his daily work recordings, he does not elaborate on the details of the event. The reader is not given any content on the nature of the "loud uproar" or what he meant by the depiction of governmental work as "God's Work." One can speculate about the potential skepticism of young engineers at the elevation of government service or about the reaction to the construction of state service and technocratic expertise as religious duty.

One can also see clear examples of a sense of religiosity that permeates Mohanakrishnan's life and work. At various points, his professional journal documents a practice of building a small temple near a newly constructed infrastructure project and efforts to visit nearby temples during his many field visits and travels in addition to his regular practice of worship. This interwoven sense of service and religious duty provided an underlying foundation to Prof. Mohanakrishnan's self-understanding and self-presentation of his life and work.

Mohanakrishnan's philosophy of service consisted of a deep-seated set of ethical and professional principles. At first glance, the foundation of his professional outlook is rooted in the Nehruvian context in which he began his career at the Public Works Department. His autobiographical representation is steeped in language that emphasizes the technical efficiency and prowess of the engineering skills being employed in the execution of infrastructural projects. Echoing Nehru's well-known characterization of dams as the new temples of India, Mohanakrishnan would present in detail the laborious construction work involved in his first major engineering project in 1948, noting that "I am detailing all these just to emphasize how methodical and steady were the preliminaries organized before taking up the great task of building the great dam" (2016a, 85). Indeed, the detailed description sheds light on the ways in which the construction of the dam was preceded by the construction of the physical infrastructure of a new community spatially stratified by class and occupation. As he would write, "The residential colony will consist of a few streets parallel, to house the junior assistants, senior assistants and up on the lines, quarters for the Supervisors. As if crowning these parallel streets, a semicircular formation was planned in which 12 SDO's [subdivisional officers] were located and at the crown, quarters for the single Execution Engineer, which was the only terraced building with two floors" (85). The organizational hierarchy of the PWD employees was intricately spatialized in the physical infrastructure of housing that was built prior to the construction of the dam. Meanwhile, the workers' housing lines "were built in rows, six units in each row, each unit consisting of a small veranda with one foot depressed roof, an all-purpose room with headroom 10 feet to the top of the roof to the left and a space for kitchen and eating behind the veranda. Six such units will be arranged in each row, the middle ones facing each other, for the residents to have social atmosphere at their level" (88). The construction of a stratified socioeconomic infrastructure thus preceded the physical construction of the dam in a way that vividly encapsulates the ways in which the Nehruvian modernist project would come to reproduce an underlying form of class stratification within the architecture of India's new interventionist state.

However, Prof. Mohanakrishnan's own social location also reveals the complexities of class formation and the role of local state-level bureaucratic employment in providing avenues for new entrants to middle-class status in the early years of independence. Mohanakrishnan describes at length his modest rural background in a small landowning family with "a few pieces of land" (2016a, 7), growing up primarily in the house of his grandfather, a village postman (7). While he was from an upper-caste Chettiar community with some landed resources, as a Telugu speaker he was also a linguistic-ethnic minority in Tamil Nadu. Decades later, even though he was a well-placed member of the bureaucracy, his attentiveness to his linguistic identity would lead him to note when he had made effective and well-delivered presentations in Tamil (a language that he was fluent in) (2016b, 9–10, 31).

Prof. Mohankrishnan's narration illuminates existing understandings of the relationship between the state and middle-class formation in a number of ways. His life story confirms the significance of the role of education in shaping middle-class formation and in the relationship between the educated middle classes and state formation in the early decades of independence (as the educated middle classes were tracked into expanding public sector employment). In keeping with this pattern, education plays a central role in Prof. Mohanakrishnan's access to middle-class status and upward mobility through state employment. He opens his autobiography with an acknowledgment of the centrality of education in shaping his life and career. As he puts it, "I am fortunate in that I have been born in a family in the Chodavaram, now Sholavaram in Ponneri Taluk of Thiruvallur District, who were anxious to give good education to the children, though they had not crossed the Elementary School stage themselves" (2016a, 3). His autobiography presents detailed discussions of his teachers, ranging from preschool to his engineering training, whose support and encouragement he honors as crucial to his advancement.

However, Mohanakrishnan's personal story also complicates existing conceptions of middle-class formation. Describing his preparation for an interview for entrance into the College of Engineering, Mohanakrishnan writes, "Till then, I had never worn a pant and a full hand shirt" (2016a, 59) and then details the process of purchasing the cloth and getting it stitched in time for the interview. This vignette, of course, speaks to the ways in which less-privileged social groups must persevere in gaining the appropriate forms of social capital that are taken for granted by the educated urban metropolitan middle classes.

Public employment in technocratic fields such as engineering in the early independence period held deeper meanings for India's middle classes, which were being shaped by new bureaucratic state structures. If education in disciplinary fields such as engineering provided public resources for individuals to gain access to middle-class status, public employment was infused with the new national ideals of technocratic developmental progress of the Nehruvian period. At the individual level, Mohanakrishnan's writings illustrate how the implementation of Nehruvian modernism through the material execution of new national goals of technical prowess and efficiency shaped the identities of state employees in distinctive ways. As prime minister, Nehru would physically and ritually embody such principles for young engineers in the public sector. After five years of work on his first engineering project, Prof. Mohanakrishnan recorded with pride the prime minister's visit to the completed dam: "Shree Nehru arrives 4PM, goes round the earth dam, all in open car, standing and wishing the people, given tea on top of the shutter House specially got ready for the function and then drives over masonry dam. When he got down on the masonry dam to walk for a short distance, I ran up to be close to him with my file of plans as part of the team" (2016a, 127). The bureaucrat's file has in recent years become a symbol of the inefficiency and sluggishness of India's public sector. Images of stacked dusty files on nondescript desks still vividly speak to the slow, low-tech pace of the state. However, for Mohanakrishnan, the file embodies a material link that connects his hidden labor of planning and construction, as he puts it, "as part of the team," to a grander vision of the state, literally and ritually embodied in Prime Minister Nehru's walk at the dam.

This sense of public purpose permeates all of Prof. Mohanakrishnan's writings, as he meticulously documents his work, ranging from the painstaking task of making hydraulic measurements to the political intricacies of negotiating interstate agreements. What stands out in Prof. Mohanakrishnan's autobiographical self-representation is a continued understanding—over a period of seventy-five years—of "public works" as a form of public service. This understanding of public sector work as public service represents an ideational and experiential space that exceeds the structural dimensions of class formation and the relationship of extraction that, as we have seen, have come to characterize India's public sector and state bureaucracy.

Prof. Mohanakrishnan's conception of service was shaped by an ethical philosophy that produced sharp critiques of the deepening networks of political patronage that infused the institutional and infrastructural decisions of the state and a set of everyday practices that endeavored to challenge or circumvent this process of politicization. Consider the following description of work culture, which he documented in 1981:

> Many officers make it a point to meet their superior officers, with sweets and presents on the New Year Day and greet them. Particularly the office bearers of the Engineers Associations go round in a group calling on the Hon'ble [Honorable] Minister, Secretaries and Chief Engineers and the Chief Engineers on their turn meet the Secretaries, the Hon'ble Ministers and so on. I had thought over the practice, I should follow, and took a decision, that I will just call on my immediate superior with a lime on hand and nothing else, as early as 1955, when I was sub divisional officer. (2016a, 310)

The practice of gift-giving in this context is suffused with the power dynamics of workplace hierarchies. What might otherwise seem like an innocuous ritual of giving sweets and presents in celebration of the New Year in this context serves as one of the naturalized practices that produce ingrained networks of patronage and supplication within organizational cultures. Mohanakrishnan's writing reveals both a keen ethnographic eye in the observation of such practices and a sustained form of resistance, over his decades of employment, to being implicated in this workplace culture of supplication.

Indeed, Mohanakrishnan's depictions of workplace practices confirm many of the public and academic narratives of patronage, personal power, and political influence that suffuse public bureaucratic institutions. Such practices range from engineers developing political connections with IAS officers (2016a, 277) to bureaucrats establishing relationships with high officials in government ministries to obtain promotions (2016a, 338) to the inclusion of lawyers for the Cauvery dispute based on "political influence"

(2016b, 44) to the personal political interventions that government officials would make to push for candidates when positions in the PWD would open up (2016a, 290). While such practices are not surprising to critics of India's bureaucratic and state practices, what is distinctive is the sustained everyday persistence that Mohanakrishnan exemplifies in his endeavor to sustain his ethical principles of workplace behavior.

The dailiness of Prof. Mohanakrishnan's ethical approach to his workplace is accompanied by both a critical analysis of and an attempt to circumvent larger structures of political patronage and influence. He documented a successful example of a circumvention of continued extension of a major canal beyond its technical capacity due to political pressure. Describing the entrenched and systemic nature of the politicization of infrastructure projects that engineers faced, he wrote,

> After the end of Congress regime in 1966, Tamil Nadu had DMK and ADMK rule alternatively every five years and with each change came more demands for extension of ayacut in their areas of followers and this was going on uncontrolled. Much later when I had to deal with their problem as the Chief Engineer (Irrigation), I sent a file with a note written in bold letters in Tamil saying this project is now committed to cater to a little more than 4 lakhs [400,000] of acres with canal extensions and so on and I see it is already bursting in its seam and would not advise even one acre more. The then Chief Minister Hon'ble M.G.R. who had great trust in my advice, rejected any further requests even from his own MLAs citing my Note. (2016a, 214)

In this example, Mohanakrishnan is successful in persuading the chief minister to resist political pressure from the MLAs (Members of the Legislative Assembly). Nevertheless, it also illustrates the immense historical weight of enduring patterns of political pressure that are exerted on government employees. The incident reveals that the expansion, or in this case overexpansion, of large infrastructural projects cannot simply be understood as the product of the technocratic outlook or disciplinary expertise of bureaucrats or engineers. Rather, elected officials, such as local MLAs, who may themselves be facing pressure from their constituents within civil society, are often at the forefront of pressing for infrastructural projects that can serve short-term electoral needs.

Mohanakrishnan's autobiographical representation is in many ways a story of a technical expert deeply committed to ideals of hard work, honesty, and pride in the design, construction, and management of water-related infrastructure that is punctuated by the everyday improprieties of patronage, power, and influence. His professional diary, which reproduces a contemporaneous daily log of events, often contains deeply felt notations when he is falsely accused of having "not cared to send a paper to" a senior official (2016b, 20) or when he must handle "rude behavior" or time wasted with "gossiping" (96) that violates his sense of professionalism. Describing one such incident early in his career, he writes of his experience with his senior colleague in the PWD,

> I had to bear his bossing, as a humble mild subordinate. He will keep calling me while in office to his room for discussion and instructions and will not allow me to dispose of the files that accumulate at my table. I am one who will not sign a file without personally reading through the current, the replies, office notes etc. Every file that is seen by me will carry my observations, made in a different ink. Once he said, "as Deputy Chief Engineer you cannot keep looking into files in office. Take them home." I was doing so. (2016a, 277)

The anecdote illustrates the everyday workplace practices that accumulate in the slow pace of bureaucratic activity that has become a much-reviled feature of the Indian state. However, Prof. Mohanakrishnan's daily entries also present detailed notes on a range of employees, engineers, and officials whom he characterizes as fulfilling high standards of hard work and honesty. It is this latter space of ethical agency within the bureaucracy that must be incorporated in any full account of bureaucratic agency.

Autobiographical works are of course performative productions of self-representation and selfhood that are not transparent reflections of a singular reality. However, this project of representation deepens rather than dislodges the significance of Mohanakrishnan's ethical philosophy and practice of public bureaucratic service. The autobiography, while published as an internal document by the PWD, is primarily targeted at an audience of employees and students specializing in water resources management and is not a publicly distributed text.[20] In addition, the second volume of the autobiography is in fact a reproduction of a contemporaneous daily diary of events and personal commentaries that Prof. Mohanakrishnan kept during his employment. The thematic focus on his ethical struggles thus presents, paradoxically, both an unfiltered and a carefully crafted narrative for future generations of employees in the PWD and in related engineering fields. Seen in this light, Prof. Mohanakrishnan's provocative statement "Government Work is God's Work" must be dislocated from a one-dimensional interpretation that could cast it as evidence of an internalization of a Nehruvian vision of technocratic progress, a blind infusion of state ideology, or a static form of "Hindu identity." The ethical space of Mohanakrishnan's life and work is shaped by but not reducible to the bureaucratic institutional and political fields that structure employment in an institution such as the PWD.

Bureaucratic Agency and Its Limits

Mohanakrishnan's history provides a highly successful case of an individual who managed to navigate through the complex institutional and political networks that at various periods prevented him from gaining promotions and positions he sought but that ultimately did not forestall his rise to the highest position within the PWD, chief engineer (irrigation). Noting that "my professional ambition is achieved," he also concludes his autobiography by underlining that he "had no God fathers. I had no special favours done to me and I stepped in, following due seniority" (2016a, 319). However, Mohanakrishnan's struggles and successes occur within the existing terrain of disciplinary, technical, and institutional norms of his field. Individuals who may seek to challenge such norms face steep constraints that are shaped by the intersecting structures produced by the dominant political and institutional fields of the water sector.

Consider, for instance, Ramaswamy Iyer's personal account of his experience in challenging some of the technocratic norms of water policy at the national level. He describes his changing relationship with the Ministry of Water Resources as he begins to question the state's technocratic approach to water management through large-scale projects, such as big dams and later on the national river-interlinking project. As he writes,

> An account of that changing relationship in capsule form would be the following: when I was Secretary, Water Resources, in the Government of India,

in the 1980s, I enjoyed a very good relationship with my colleagues and subordinates; that goodwill continued for a while after my retirement, but changed to strong disapproval as I began questioning and criticising big-dam projects; the disapproval reached a peak in the years 1998–2005; then slowly, over a period of time, anger against me mellowed, and the broken relationship was partially mended—but only partially; some embers of the old uneasiness still remain and can ignite easily. (2013, 168)

Iyer describes in detail how in the 1990s, after his retirement from formal government service, his growing receptiveness to environmental criticisms of large dams, such as the highly contested Sardar Sarovar Project on the Narmada River and the Tehri Hydroelectric Project, angered what he classifies as the "water establishment."

Iyer explains how his appointment by the Ministry of Water Resources to a "Five Member Group" and by the Ministry of Power to an expert committee to evaluate the environmental and displacement effects of the projects led him to a deep intellectual reconsideration of the costs and benefits of such large-scale projects that have been supported by the central government. He describes at length how his break from the dominant norms of the institutional fields produced significant resistance through the professional and organizational networks that structure such fields. As he writes,

> My relationships with former friends and colleagues came under a strain. I was no longer welcome in my old Ministry. I used to visit the Ministry occasionally to meet people and get myself briefed on developments of interest, but this became increasingly difficult. Senior officials did not want to meet me. The Water Establishment's disapproval of me was even greater than its disapproval of Medha Patkar. After all, she was the Enemy, but I had been part of the Water Establishment a few years earlier. In the eyes of the Establishment I was one of them. It was as a former Secretary Water Resources that I was nominated to various government committees and commissions, and there was dismay in official circles when my thinking changed and I began speaking a different language. The dismay changed to anger. I was regarded as a renegade who had deserted the ranks and joined the enemy camp. People in the Ministry and in the Central Water Commission (particularly engineers) who had earlier been well disposed towards me became cold. Some former colleagues

> who continued to maintain friendly relations with me had to contend with the disapproval of their engineering brotherhood. (2013, 171)[21]

Iyer's brief but incisive personal account of his experience with the national "water establishment" points to the significant challenges of expanding an ethical challenge to dominant institutional networks and practices to address the more fundamental disciplinary and technocratic modes of power that shape large infrastructure projects. As he notes, the institutional and personal hostility subsided as the public contestation over the Narmada Sardar Sarovar project began to abate (and since the final intervention of the Supreme Court in effect sanctified the project). However, this antagonism arose once more with his sharp critiques of the national river interlinking project that is now underway with substantive state support. As he puts it, "The engineering establishment set much store by the project and were made angry by my criticisms. The dormant official disapproval of me became alive and active" (2013, 174).

Iyer's personal account provides an illuminating understanding of the ways in which state power, institutional practices, professional networks, and disciplinary norms intersect and constrain the kinds of criticisms and challenges that can take place within bureaucratic structures. In this context, the disciplinary norms and professional networks of engineers are not simply determined in a simplistic way by state censorship or prohibition but are shaped by the weight of historical legacies that have connected the state, infrastructural projects, and the disciplinary projects of bureaucratic institutions that create, support, and manage such projects. As Iyer insightfully argues, since the colonial period,

> the engineering profession has commanded great respect, and it has been customary to talk in reverent tones about "great engineers" or "eminent engineers." The tradition established by [colonial British engineer] Cotton and others has been absorbed by successive generations of Indian engineering students. A certain professional pride and a sense that they are pursuing a socially useful profession has been inculcated in them, and quite rightly so. For a century and a half it has been taken for granted that it is good to build dams for irrigation or for the generation of electric power. Against that background, it must have been extremely disorienting for the profession to be told that dams are not necessarily benign, that they could do a great deal of

harm, and that dams must not be built unless they are unavoidably necessary. A highly respected profession which had taken for granted its value to society suddenly found its self-esteem undermined. (2013, 175)

The historical significance of engineering has been deeply embedded in dominant societal norms. In an evocative cultural analysis of poetic representations in Tamil Nadu, Anand Pandian has called attention to the ways in which communities in Madurai that benefited from infrastructural works of the colonial hydraulic engineer John Pennycuick have honored his achievements through a long history of popular cultural memorials expressed through song and poetry.

Pandian notes, in the midst of a rapidly changing liberalizing India,

> In December 2001, a few young men from the bustling town of Cumbum [in southern Tamil Nadu] circulated invitations to an opening gala for the new internet café they had just established in the busy market. . . . The invitations proudly stated that their Green Valley Internet Browsing Centre was dedicated to the memory of "Respected Benny Quicc, The Founder Cumbum Green Valley." This phonetic rendition of a foreign name may have been slightly off the mark, but the historical sentiment was unmistakeable. The browsing centre was inaugurated in the name of Colonel John Pennycuick, the colonial hydraulic engineer almost universally credited today with having brought a perennial stream of river water into the Cumbum Valley and the arid plains of Madurai. (2003, 12)

This historical memory that codifies the discipline of engineering with honor and social status intersects with a long history in which higher education in fields such as engineering has played a significant role in the formation of India's educated middle classes (Fernandes 2006).

It is this sense of historical pride and self-esteem, as we have seen, that continues to serve as a legacy for engineers within institutions such as the PWD, even as newer institutions such as Metrowater have begun to take on more significant public roles in the water sector. Ramaswamy Iyer's personal account is a useful cautionary reminder of the deep structures of the institutional fields that shape and constrain the agency of employees in such organizations. Iyer's enactment of his bureaucratic conscience faces steep forms of institutional and disciplinary resistance that are in turn shaped by the weight of intersecting modes of power of colonial, postcolonial, and postliberalization state practices.

Abuses of state power, including widespread corruption and rent-seeking behavior, are a significant material dimension of the bureaucracy. However, both ideological critics of development and advocates of reform too often tend to operate with unidimensional conceptions of bureaucrats as corrupt, inefficient, or homogenized rational actors. Such conceptions do not fully capture processes of class formation and state-class relations that are entangled in the production of the bureaucratic workforce or the complex forms of agency and subjectivity of bureaucrats who must navigate a difficult political and institutional terrain. The complexity of the figure of the bureaucrat in effect stems from the ways in which bureaucrats are in many ways the human embodiment of this unwieldy boundary between the state and civil society. This wider perspective is critical for any understanding of questions of effective and accountable governance.

Conclusion

IN JUNE 2019, CHENNAI GAINED GLOBAL NOTORIETY AS ITS MAJOR reservoirs were depleted by an extended drought, and international media narratives fed on the spectacle of one of India's major metropolitan cities going dry. Local opposition party politicians mobilized protests, and trains brought in emergency water. Inequalities embedded in access to water resources through private markets were brought to the surface, and more-nuanced stories of the crisis pointed to deeper problems with urban development and governance. The crisis even overwhelmed more-privileged water consumers, who had usually been able to maintain access to water through private markets. As businesses and industries struggled to deal with the crisis, they began to face the structural strains on water that have been intensified by urbanization and development in the postliberalization period.[1]

If the 2019 drought brought to the fore the strains on urban governance in periods of water scarcity and the deeper relations of extraction with rural areas (as the supply of groundwater to the city was intensified), the monsoon season of the preceding year laid bare a competing set of pressures on the governance of water in Tamil Nadu and its neighboring states in southern India. The generous rains in 2018 produced surplus water that filled the Cauvery's catchment areas and Tamil Nadu's Mettur Dam, which supplies water for the state's agricultural areas. The bounty of surplus waters produced a season of relief from the tense standoff between Tamil Nadu and Karnataka over the sharing of the Cauvery waters. With the dam's water level reaching its full capacity for the first time after five years, an "exultant" chief engineer of the PWD's Water Resources Organisation indicated that all of the excess water was being let out for irrigation.[2] Meanwhile, exuberant reports on the monsoon's gift also noted that water from the Mettur Dam would reach Veeranam Lake, which had been harnessed to provide water supplies for Chennai.

Within a few weeks, the promise of the euphoric media narratives began to give way to the fissures of the political, economic, and institutional challenges that have constrained the governance of water. The lack of proper institutional maintenance of water bodies in the area (for instance through desilting, a responsibility of the PWD) meant that the water bodies had not been filled by the surplus Cauvery waters.[3] Farmers argued that illegal sand mining was playing a significant role in preventing the replenishment of water bodies.[4] While these institutional failures became mired in the politicization of water management, with charges and countercharges by the opposition and ruling political party, the heavy monsoons soon set into motion the too-familiar oscillation between the perils of scarcity and floods. While areas in the vicinity of the Cauvery were soon overwhelmed by flood warnings, unprecedented floods in neighboring Kerala overwhelmed the state and once again, reigniting the Mullaperiyar Dam conflict. Meanwhile, within days of simultaneous stories of flooded areas in various districts of Tamil Nadu, reports surfaced that Metrowater had begun plans to tap groundwater again to meet dwindling water supplies for Chennai. While scanty rains in the city were reported as the source of dwindling supplies, the oscillating stories of floods and drought highlight the deeper, systemic factors that have obstructed the effective management of water resources and that lie beneath media stories of reservoirs going dry. The workings of institutions, in both their mundane and weighty forms, do not have the glamor of stories of crisis or the poetic flavor of stories of subaltern communities. Yet they are the heart of democratic governance and are a crucial site for understanding how inequalities are produced and reproduced.

The governance of water both illuminates the nature of state power and has itself been shaped by the remaking of the state in postliberalization India. The nature of water is such that it crosses the borders of territorialized administrative structures, the categorical separation of various sectors of the economy (such as industry and agriculture), the distinctions between instrumentalist uses of a natural resource for the economy and the human needs for life and survival, the spatial distinctions between localities (such as cities and villages), and the temporal divisions of historical periods (of colonialism, twentieth-century developmentalism, and twenty-first-century reforms).

The significance of water for both specific economic uses and the universal needs of human life means that the governance of water brings to the fore the underlying contradictions, power relations, and contestations that are embedded in democratic institutions. The question of the governance of water is also both a distinctive and an archetypal case for a broader understanding of the liberalizing state. In contrast to the methodological biases that stem from analyses of particular sectors of the economy (for instance new economy sectors such as informational technology or pharmaceuticals), water cuts across all sectors of the economy. Policies of liberalization are transforming the governance of water, and the governance of water is fundamentally intertwined with and illustrates the nature of the remaking of the political economy of the liberalizing state.

While the era of economic liberalization is conventionally associated with the principles of decentralization and privatization, such principles are embedded in broader institutional and policy frameworks that consolidate and intensify centralized state power. This form of centralized state power has both built on earlier forms of centralized state authority associated with the colonial and developmental state and set into motion new forms of centralized authority that are distinctive to the postreform era. As we have seen, such forms of authority are concentrated within city-centric modes of urban governance. These forms of centralized authority both reproduce and intensify inequalities between and within metropolitan city centers on the one hand and small towns and rural areas on the other hand. This account of the city of Chennai is in fact an account of broader local, regional, and national socioeconomic relationships that are unfolding in the context of global ideational and policy approaches to water.

An adequate understanding of the nature of this centralized state authority necessitates a conceptual shift in conventional studies of reform policies in India. The two bedrock principles of reform that have broadly governed India's project of liberalization since the 1990s are privatization and decentralization. Such principles, which have been entrenched in public rhetoric, institutional frameworks, and economic policies across various sectors, contain within them a presumption of a shift in the role of the state. In an idealized scenario, this shift would be encapsulated as one in which the Indian state transitions from the command-oriented developmental state to a regulatory state that facilitates rather than controls economic activities. The case of water reforms has instead shed light on a set of contradictory processes in which policies of reform have reoriented as well as continued and intensified long-standing practices and modes of centralized state power.

The point at hand is not that decentralization and privatization are not important or significant processes that have been taking root in contemporary India. However, the persistence and consolidation of centralized authority is not purely an aberrant legacy of the older model of India's command state or a byproduct of a corrupt bureaucracy. Rather, the dynamics of state centralization are an inherent framework of the postliberalization model of development that has become dominant in the twenty-first century. Exceptionalist arguments that posit that reforms that are designed to decentralize state authority in India are obstructed purely by the conditions of India's political or institutional fields do not account for the ways in which institutional and economic reforms in fact produce the very frameworks and configurations of concentrated state authority that such reforms purport to transform.

Institutional reforms designed to scale back the role of the state through processes of decentralization and the participation of private sector actors have in fact produced a redistribution of centralized institutional power rather than a shift from centralized to decentralized state governance. An adequate understanding of this kind of remaking of the postliberalization state through the redistribution of institutional power has necessitated an analytical framework that develops a more nuanced understanding of the workings of bureaucratic organizations. From such a perspective, processes of reform can be understood in the ways in which they redistribute power and resources both within the bureaucratic field and within civil society. At one level, such an approach provides the conceptual space to address the complex relationships between the state and civil society that shape the orientation and effects of bureaucratic institutions (Evans and Heller 2015). At another level, such an approach provides the analytical space for an understanding of the ways in which policies of reform can produce new forms of centralization within some arenas of the institutional field of the water bureaucracy, even as they weaken or curtail the authority of other institutional sites. Inequalities are both produced and intensified by such institutional reforms. The redistribution of institutional power, for instance, intensifies the power of wealthier communities in metropolitan cities while producing weaker governing bodies in rural locales and small towns. Regulatory reform, in the process, is transformed into a form of regulatory extraction. The resurgence of state authority in ways that are both familiar extensions of historical legacies and new attributes of a liberalizing state raises the question of where the role of private capital lies in this remaking of the state.

Governance and the Debate on Privatization

In recent years, critics of economic reforms have called attention to the problems of "neoliberalism" and the political and economic power of private capital. Indeed, the centrality of private capital is built into the focus on economic reforms. However, critics of neoliberalism have often underestimated the role of the state (Fernandes 2018a). The centrality of the state has been intensified by the shift from the 1980s Washington Consensus, which emphasized the retreat of the government, to the post–Washington Consensus, which has recentered the state in global models of institutional reform. The arguments of this book thus have crucial insights for comparative contexts. As we have seen, the state remains the central actor in shaping the distribution of resources, and processes of privatization have been woven into the remaking of the liberalizing state. Take, for example, the World Bank's broader shift toward ensuring that its financial investments are carefully structured within clear frameworks of state governmental institutions and accountability.

Processes of privatization mirror the dynamics of centralization and decentralization—they have left sites of centralized state power intact while targeting weaker sites, such as ULBs, for programs of privatization. In the case of Chennai, for example, privatization has largely kept in place the centralized authority of bureaucratic organizations such as Metrowater and the Public Works Department. In keeping with national trends, the systemic processes of privatization that have taken place have focused primarily on institutional restructuring that has downsized workforces and increased the use of subcontracting. Furthermore, as the Chennai case illustrates, privatization does not necessarily weaken or displace centralized state authority. Such dynamics raise the question of what form privatization takes. For instance, as journalist Nagesh Prabhu has noted in the case of the controversy over the attempted privatization of the Delhi Jal [water] Board, "Assets, staff, revenues, and tariff-setting" would have remained under governmental control (2017, 267). The point of relevance for our purposes is not an ideological one on the pros and cons of privatization but a substantive and analytical one. State authority is reconsolidated by privatization.

We have seen that institutional and economic reforms have not undermined but have redistributed institutional power. This disaggregation of the state is necessary to adequately understand *how* the relationship between the state and private capital plays out within particular sites and sectors of the economy. Scholars have now long since been preoccupied with the contested nature of the boundary between the state and civil society. The questions of power and inequality that such scholars are contending with requires a deeper understanding of the boundaries *within* the state, which are codified through institutional structures. A relational perspective on institutions allows us to ask and understand the ways in which the ascendency and decline of particular institutions both reflect and produce inequalities and relationships of power. Modes of governance in this context become a central means for the systemic reproduction of inequalities within and between various locales.

Consider, for instance, one of the central ways in which privatization has taken root in Chennai. Privatization has often been a consequence of a retreat of the state due to incapacities or willful action rather than to conscious policies of privatization. The failure of the state to provide services and to develop effective regulatory frameworks of governance has in such cases provided a space that has subsequently been occupied by private actors, privatized practices, and illegal informal networks and organizations. As we have seen in the case of the emergence of private water markets in and around Chennai, such water markets are the product of both state incapacities (such as the expansion of the water tanker and water bottle industry in the face of inadequate water supplies) and state intervention (such as the state's promotion of groundwater extraction to compensate for the absence of water resources in times of drought).

This does not of course mean that the direct impact of private companies on water resources is not significant. High-profile social movements have, for instance, successfully targeted private companies for the damage they have caused to both water resources and the livelihoods of poor communities. The high-profile case of local tribal and rural grassroots protests successfully pressuring Coca-Cola to close its factory in Kerala is a well-known example of such movements.[5] The movement that sought to combat both the depletion of water resources through the extraction of groundwater and pollution caused by the company's operations represents one of the most visible examples of the direct negative effects of private corporations. In Tamil Nadu, a major protest movement against Vedanta's Sterlite Copper plant because of widespread pollution and health issues was successful in pressuring the Tamil Nadu government to close the plant.[6] In these high-profile cases, the question of state authority remained a foundational element, as protestors and civil society organizations had to pressure governmental officials and work through the various levels of the court system. In the Vedanta case, the impact of centralized state authority was profoundly demonstrated in police shootings of protestors. Such severe examples underline the importance of understanding how the state is acting or not acting in its governance of water. These examples of securitized state action are not isolated excesses—they are on one end of a continuum of the range of centralized actions of the postliberalization state. The nature of centralized state authority is a topic that is rich with questions for future research about democratic accountability that are of relevance to the myriad issues, conflicts, and concerns that make up the field of water governance.

Governing Water and the Question of Bureaucrats

In the midst of these serious challenges for the governance of water, there is perhaps no more reviled and caricatured figure in contemporary India (and in comparative contexts) than the bureaucrat. The weight of corruption and the inability of local bureaucracies to maintain infrastructural services are very real and heavy burdens on citizens. Indeed, local water bureaucracies and utilities such as the PWD and Metrowater are themselves publicly distrusted institutions. Yet, as we have seen in this volume, a close reading of documents and interviews with bureaucrats yields a more nuanced picture of bureaucrats and the agency that they do—or do not—exert.

Historical and sociological dimensions of institutional practice reflect the broad patterns of institutional practice and the macro structures that condition and constrain such practices in the water bureaucracy. Within the contours of these bureaucratic fields, state employees must negotiate the formal institutional rules, informal cultures, and political dynamics of their organizations. A move away from a focus on a generalized, static form of corruption is needed not because corruption is not real and an often overwhelming part of both everyday and institutional life in India. However, a singular focus on corruption prevents us from gaining a deeper understanding of what is and is not working within state bureaucracies. Public institutions matter within the framework of a democratic polity, and it is as crucial to understand the spaces in which bureaucrats are trying to effectively perform their duties as it is to draw attention to dysfunctional or corrupt practices. This book has thus sought to create the analytical space for a more nuanced understanding of bureaucratic agency, which can delineate when bureaucrats and their organizations are constrained by the structural limitations of their political and economic environments, when they are engaging in corrupt practices and when they are attempting to navigate complex institutional fields in order to perform their regulatory duties.

What emerges is not an image of the bureaucracy that is free of corruption and inefficiency but one in which there are strong pressures from political, economic, and internal organizational constraints that deter or suppress the actions and insights of those bureaucrats who are indeed committed to their professional and institutional duties. A more complex conception of bureaucratic agency exceeds conventional accounts of corrupt, inefficient, or rational actor typologies. The more expansive discussion of bureaucratic agency has included examples of local bureaucratic agency that has enhanced interstate cooperation and attempted to carry out regulatory functions. A fuller understanding of bureaucrats also takes into account the question of ethical agency, which is often not associated with the figure of the bureaucrat.

The question of ethics does not of course in itself preclude deep-seated problems associated with the exercise of state power. The realm of ethics is a field that encodes relationships of power and ideological predispositions. Bureaucracies, as many scholars have shown, are enmeshed in the modernist ideologies that are associated with the dominant ideals of nation-states. Nevertheless, the case of Mohanakrishnan provides a more nuanced sense of the complex subjectivity of bureaucrats. Bureaucracies are, after all, still largely understood at best in Weberian terms of rationality or efficiency or at worst as institutions of organized indifference (Herzfeld 1992). Opening up the analytical space for an understanding of the affective dimensions that shape the subjectivity of bureaucrats and the potential for ethical agency is a crucial dimension of the question of governance.

Critics of development and, more recently, of neoliberalism often rightly point to the problematic reification of technical and professional "expertise" (Laurie and Bondi 2005). However, global challenges such as climate change and health pandemics remind us of the critical need for accountable and inclusive models of governance. The scale of such crises also means that responses also require a scale of governance that cannot simply be delegated to local, decentralized organizations.

In the case of the politics of water, the likely threat that floods and drought due to climate change are accelerated and that environmentally unsound urban development will intensify and expand in scale means that governmental responses will also need to operate in systemic ways. Such responses of course need to be shaped by the knowledge and needs of local communities. However, the romanticization of local grassroots approaches can itself operate as an offshoot of the neoliberal imaginary (Hall and Lamont 2013).

Governance, then, cannot effectively work without developing the means for reforming bureaucracies and opening up the institutional and political space for bureaucrats who do bring with them a sense of personal ethics. This space is foreclosed because of the ways in which graft and patronage become part of the internal reward structure of bureaucracies and the internal, informal punitive structure of organizations that marginalize or penalize those employees who do not consent to the patronage politics of institutional cultures. Not surprisingly, such structures produce ineffective governance. This is compounded by the very reform processes that have led to cutbacks on the bureaucratic workforces—cutbacks that are made as these workforces must manage the expanding consumer demands that strain water resources and infrastructure. As my research has noted, to revisit one small example, the water bureaucracies in Tamil Nadu do not have the labor power to consistently monitor water bodies and infrastructure in the state. The deteriorating state of water bodies then comes to the fore during crises of floods or droughts. If public institutions are to effectively work as *public* institutions that are not slanted toward networks of patronage or constituencies with political and socioeconomic power, debates on governance will need more-sustained analyses and approaches to public employees of such institutions.

The Comparative Implications of India's Water Reforms

The remaking of state authority through reforms in India's water institutions holds important implications for our understanding of global trends in the governance of water. One of the biggest challenges for water governance is the impact of urbanization. In the first global survey of water sources for large cities, McDonald et al. estimate that "one in four cities, containing $4.8 ± 0.7 trillion in economic activity, remain water stressed" (2014, 96). Given the economic and political power of cities, urban governance also remains a critical site for the concentration of state authority in comparative contexts. This nexus means that the governance of water will remain both a fraught site as urbanization places new stresses on water resources and a key arena for the exercise of state authority. Indeed, Tamil Nadu's complex interstate water negotiations and disputes are illustrative of a broader comparative pattern of political contestations over shared river basins (Moore 2018). Understanding how states manage this nexus of urbanization and water stress is of broad import in a globalizing world that continues to advance economic policies that expand urbanization in the Global South and deepen the potential stresses on water resources.

Contemporary processes of urbanization and their corresponding stresses on water must be contextualized in an expansive temporal and spatial framework. Both the modes of governance and the challenges that states face are shaped in complex ways by the historically produced political-economic structures that shape and constrain water bureaucracies. In the case of postcolonial contexts, such historical processes are shaped in distinctive ways by the historical legacies of both the colonial state and the developmental agendas of the postcolonial state. Urban governance is fundamentally enmeshed in rural-urban, regional and national, and political and economic processes. The story of the city, from this perspective, must be understood as a reconfiguration of power across and within these spatial scales.

Most significantly, these temporal and spatialized complexities call for ongoing research agendas that address the underlying connections between institutions tasked with the management of water resources at various spatial sites and scales. One of the major global trends in recent years has been the shift to policies and principles of governance based on the model of integrated water management. Indeed, the promotion of ideals of water management through integrated approaches that address the complex configurations of land, resources, and environmental sustainability are worthy goals. However, as the case of India's water institutions has illustrated, the governance of water is compartmentalized through discrete and often competing institutions, and these institutional silos mirror the presumed territorialized divisions between cities, towns, and villages and between local and central governmental institutions. Institutional reforms, in this context, are both shaped by and are the means for the production of underlying reconfigurations of power and inequality. Further work is needed to develop integrated institutional analyses in comparative contexts that can grapple with the relational power-laden dynamics of institutions tasked with the governance of water. Such a relational understanding is particularly crucial for scholars and practitioners who are concerned with questions of inequality and the water needs of poorer urban and rural communities. Such a perspective, for instance, asks how models of decentralized governance may inadvertently be part of a set of institutional mechanisms that are consolidating centralized state authority over water. An analysis of this relational institutional field does not seek to dismiss the value and significance of community-based models of governance. On the contrary, it is of particular import for advocates of decentralized or community-based governance to grapple with the implications of differential forms of institutional power.

At the broadest level, this unsettling of the institutional silos of governance also calls for an unsettling of the demarcation between state institutions tasked with economic reforms and development on the one hand and those tasked with the governance of water on the other. The current strains on water governance stem from a structural and institutional disjuncture within the policies of reform that nation-states such as India have been implementing. Institutional reforms of water governance have been treated as a closed system that is demarcated from policies of investment, land use, and urban development that are fundamentally intertwined with the governance of water. This kind of institutional segregation produces a profound contradiction for water bureaucracies that cannot be explained away by conventional accounts of corruption and inefficiency. In the Global South, where economic pressures drive competition for private investment and where urbanization is a natural corollary of economic growth, the underlying structural stresses on water will continue to place such strains on governance.

Water Governance and the Crises of Climate Change and Global Inequality

The governance of water is now facing major global crises of climate change and global inequality. The arguments and research of this book have important implications for these pressing global challenges. On the one hand, we have seen the ways in which institutions play a critical role in the reproduction of inequality and the creation of mechanisms of extraction that produce new inequities of access to water. On the other hand, climate change has already begun to intensify cycles of droughts and floods. These twin crises also, of course, intersect, as the impact of climate change has acute effects on marginalized communities and poorer countries. Governments and international organizations are focused on needed macro policy responses. Meanwhile, grassroots communities struggle to foreground critical questions of environmental justice.

This book illustrates that both the problems of and solutions for climate change require a deeper understanding of the mechanics of state bureaucracies. The mechanics of governance on the ground, as we have seen, are a long distance from idealized global policy models and norms. Global policy models (as with the case of models of institutional reform) are often developed in abstracted forms, without an understanding of the local and national political, social, and institutional contexts and constraints in particular places. Yet such policies are implemented by local bureaucrats within local institutions. Moreover, the scale of problems associated with climate change requires broader forms of effective and accountable governance.

Finally, one of the critical implications of this book is the need to recenter the role of global models of economic reform in the exacerbation of global challenges of climate change. As we have seen, the challenges of governing water in India are rooted in the long-term effects of successive models of development by the colonial, developmental, and postliberalization state. In global policy and public discourses, there is often a decoupling of the discussions of growth-oriented economic reforms on the one hand and climate change on the other. However, as we have seen, intense forms of water scarcity and problems of flooding have been intensified and often caused by unplanned urbanization, which is in turn directly produced by the dominant model of economic reforms. Responses to climate change and the environment will remain inadequate if they do not simultaneously rethink dominant global economic policies. As we have seen, local water bureaucrats struggle to manage resources because urban planning (shaped by private investment) is placed in a different institutional silo. Water bureaucrats are in effect tasked with managing a situation in which they have no authority over the causes of water scarcity. This is a microcosm of the institutional siloing of global economic policies of reform on the one hand and global institutions focused on climate change and the environment on the other.

Water compels us to confront the complex historically produced configurations of state power, politics, land, infrastructure, developmental and economic policies, and human life. Such complexities that shape the governance of water are particularly fraught in the context of climate change. This book cautions us that the scale and potentially catastrophic effects of climate change paradoxically require us to turn away from the spectacle of such effects to the everyday, mundane practices of organizations and institutions that implement policies on the ground. The book's analysis of such institutional practices—steeped in the weight of historical, political, and social contexts of particular places—also caution us to move away from sanitized policy responses and modular social science approaches, which are too often dissociated from the everyday realities of governance. The challenges of governing water call on us to think about the kinds of located, effective institutional responses that are critical in these times of change and crisis.

NOTES

INTRODUCTION

Epigraphs: "Use MGNREGA Funds for Water Conservation—Modi," *Hindu*, April 24, 2018, www.thehindu.com/news/national/use-mgnrega-funds-for-water-conservation-modi/article23659940.ece. "2019 Water India Expo," Smart Cities Mission, New Delhi, India, accessed November 29, 2021, www.waterindia.com/about-us.aspx.

1. For a public example of this, consider, for instance, the iconic film *Mother India*.
2. For instance, India's economic growth has been largely spurred on by the services sector and industries such as IT, while traditional sectors, such as agriculture and manufacturing, have not fared as well.
3. Within the policy realm, the shift toward local governance was built into specific legislative frameworks that were designed to promote decentralization. The 1992 seventy-third and seventy-fourth constitutional amendments were specifically designed to strengthen both rural and urban governance in villages and small towns. Decentralization has had varying implications for restructuring state authority in postliberalization India. Such variations have been shaped by a range of factors, including the nature of local state governments, local political contexts, societal dynamics and elite capture, and the institutional capacity of local governments (see Manor 2016; S. Singh 2016; Singh and Sharma 2007).
4. Govind Gopakumar (2012), for instance, provides a rare and important analysis of the ways in which varying domestic political coalitions have shaped the trajectory of water reforms in urban cities in India.
5. There is also a distinctive stream of scholarship that is focused on the impact of states on social welfare rather than purely on the nature of reforms and investment (see Deshpande, Kailash, and Tillin 2017; P. Singh 2016).

6. Water is listed as entry 17 on the State List. "Water Information," Central Water Commission, Ministry of Jal Shakti, Department of Water Resources, River Development and Ganga Rejuvenation, Government of India, accessed November 29, 2021, www.cwc.gov.in/water-info.
7. This paradoxical nature of the Indian state is not new. One of the key features of the Indian state that scholars have long grappled with is the contradictory nature of what Lloyd and Susanne Rudolph called the paradox of the "weak-strong state" (1987).
8. See, for example, Aman Sethi, "At the Mercy of the Water Mafia," *Foreign Policy*, July 17, 2015, https://foreignpolicy.com/2015/07/17/at-the-mercy-of-the-water-mafia-india-delhi-tanker-gang-scarcity.
9. I use the name Tamil Nadu in accordance with governmental and public conventions; note that the culturally specific linguistic name of the state is Tamilnadu.
10. The parties emerged out of the Dravidian movement in the state, which centered on conceptions of justice and equality that focused on caste hegemony (specifically Brahmans) and caste discrimination as well as cultural nationalist conceptions of ethnic-linguistic Tamil identities.
11. "Level of Urbanisation," Ministry of Housing and Urban Affairs, Government of India, accessed November 29, 2021, https://mohua.gov.in/cms/level-of-urbanisation.php.
12. For a useful discussion of such organizations in Mumbai, see Anand (2017). For an example of the critique of privatization, see Urs and Whittell (2009).

1. FORMATION OF INDIA'S WATER BUREAUCRACY

1. Patel would famously passionately defend the Administrative Service as India in his speech to Parliament on October 10, 1949 (see Singh 2017, 247). This has become known in Indian public culture as Patel's defense of the IAS as India's "steel frame."
2. Precolonial forms of social and political power were also fundamentally linked to the control and distribution of water resources (see Ludden 1985).
3. For a rich discussion of practical expertise in flood-prone Orissa and the deleterious effect of modern colonial technical projects, see D'Souza 2006.
4. For a more extensive discussion of the formation of the universalized science of irrigation that was taking root, see Gilmartin 1994 and Mosse 1999.
5. On this contrast, see Mosse 1999, 310.
6. Peter Mollinga, for instance, has argued that famine protection was a real objective. See his discussion of the famine commissions in 1880 and 1901–3 in Mollinga 2003.
7. Such laws included the 1873 Northern India Canal and Drainage Act, the 1882 Easement Act on groundwater, the 1920 United Provinces Minor Irrigation Works Act, the 1931 Madhya Pradesh Irrigation Act, and the 1935 Government of India Act that gave provinces rights over water.

8   The legal structure, as Philippe Cullet has noted, produced a significant linkage between land and water. The control of groundwater was connected to control of land. The result was that property rights developed around the ownership of water (2009, 28).
9   Sadr-ul-Mahan Political Department, His Exalted Highness the Nizam's Government, Hyderabad-Deccan, to the Secretary to the Honourable the Resident at Hyderabad-Deccan, December 26, 1935, no. 3186.
10 Sadr-ul-Mahan to the Secretary.
11 PWD memorandum no. 2450-D/36-2, July 2, 1936. The conference took place on July 23, 1934.
12 PWD memorandum, 28.
13 This also extended to the PWD's interpretation of riparian law.
14 "Water in Indian Constitution," Ministry of Jal Shakti Department of Water Resources, River Development and Ganga Rejuvenation, Government of India, accessed December 4, 2021, www.mowr.gov.in/water-indian-constitution.
15 For critical work on development and water politics, see Baviskar 2005; Agarwal and Narain 1997. For a discussion of this developmental model of political economy in the early decades of independence, see Frankel 2015; see also Gupta 1998. There is a very large amount of scholarship on dams in India. For a good overview of the literature, see Joy et al. 2008.
16 The weight of such institutional continuities that shaped the state's approach to water in postindependence India does not of course mean that there were no significant changes. In the context of critiques of India's developmental state, it is often too easy to forget the shift in state objectives. Thus, for instance, food security and self-sufficiency were central objects of the state in the early decades of independence.
17 The formal name of the ministry has undergone changes over time. The current name is Ministry of Jal Shakti. For the reader's accessibility, I use the abbreviated name Ministry of Water Resources throughout the book.
18 Other significant water-related issues, such as governance over droughts and floods, were treated by distinct institutional structures and efforts. On problems with drought governance, see Jairath 2008.
19 See a similar discussion of the PWD in UP in Gould 2011.

2. THE REGULATORY WATER STATE

1 This has ranged from political discourses such as Arvind Kejriwal's (then activist who would later become Delhi's chief minister) campaign against the privatization of water in Delhi to NGO activities in Mumbai (see Anand 2017) to academic works (Shiva 2016).
2 The classic case of expanded privatization is Chile.
3 See, for example, the creation of specific procedures for such initiatives and the establishment of a PPP Approval Committee, F.No.2/10/2004-INF, Government of India Ministry of Finance Department of Economic Affairs, 2005.

4   Such centralizing tendencies are not limited to the water sector. For example, while the planning commission was disbanded, it was replaced by the NITI Aayog, which has been placed under the centralized authority of the prime minister. See Swenden and Saxena 2017.
5   The Constitution (Seventy-Third Amendment) Act, 1992, Government of India, accessed December 4, 2021, www.india.gov.in/my-government/constitution-india/amendments/constitution-india-seventy-third-amendment-act-1992.
6   The Sukthankar Committee recommended devolving responsibility to neighborhood/resident associations. The Ministry of Urban Development formulated the Pooled Finance Development Fund Guidelines for small and medium ULBs (see Hoque 2012)
7   The World Bank has been advocating this measure in the irrigation sector since 1998. See Koonan and Bhullar 2012 on this point and for an assessment of the WRA model.
8   Government of India Notification F.No.2/10/2004-INF, Government of India Ministry of Finance Department of Economic Affairs, November 29, 2005.
9   Thus, for example, all interlinking projects are to be classified as "national projects" in order to speed up funding. Vishwa Mohan, "Linking of Rivers May Get National Tag," *Times of India*, https://timesofindia.indiatimes.com/india/govt-may-declare-inter-state-river-linking-projects-as-national-projects/articleshow/62544432.cms.
10 For a critical discussion, see Iyer 2012.
11 "Jawaharlal Nehru National Urban Renewal Mission," Ministry of Housing and Urban Affairs, Government of India, accessed December 4, 2021, https://mohua.gov.in/cms/jawaharlal-nehru-national-urban-renewal-mission.php.
12 "Jawaharlal Nehru National Urban Renewal Mission."
13 ICICI Bank Limited (formerly ICICI Ltd), Housing Development Finance Corporation Limited and Infrastructure Leasing and Financial Services Limited (see TNUDF, n.d.).
14 These reforms were implemented in Tamil Nadu in 1994 (GTN 1994).
15 The act was amended in 2002 (GTN 2002).
16 Networking of Rivers vs. IN RE, October 31, 2002, Supreme Court of India, Casemine, accessed December 4, 2021, www.casemine.com/judgement/in/56ea95bf607dba382a0794c2.

3. FEDERALISM AND INTERSTATE NEGOTIATIONS

1   See, for example, Tamil Nadu's attempt to compete with Andhra Pradesh for investment in the IT sector (Kennedy 2004).
2   The amendments still contain loopholes that allow delays. For instance, there is no time limit provided for publication of the tribunal decision in a gazette, which is a legal requirement for the implementation of an award (see Iyer 2002). I use Cauvery Tribunal or Cauvery Water Tribunal as abbreviated names for the full formal name, Cauvery Water Disputes Tribunal.

For the original act, see "The Inter-state River Water Disputes Act, 1956," Government of India, accessed December 5, 2021, www.indiacode.nic.in/bitstream/123456789/1664/3/A1956-33.pdf; for the text of the 2002 amendment, see "The Inter-state Water Disputes (Amendment) Act, 2002," Indian Kanoon, accessed December 5, 2021, https://indiankanoon.org/doc/1048477. For the 2019 amendment, see "The Inter-state River Water Disputes (Amendment) Bill, 2019," PRS Legislative Research, accessed December 5, 2021, https://prsindia.org/billtrack/the-inter-state-river-water-disputes-amendment-bill-2019.

3 "Cauwery [sic] Water Row: Inter-state Traffic Comes to a Standstill at Border," *Deccan Chronicle*, September 8, 2016, www.deccanchronicle.com/nation/current-affairs/080916/cauwery-water-row-inter-state-traffic-comes-to-a-standstill-at-border.html; "Cauwery [sic] Water Row: Water Dispute Turns Violent," *Deccan Chronicle*, September 13, 2016, www.deccanchronicle.com/nation/in-other-news/130916/cauvery-water-row-water-dispute-turns-violent.html.

4 See, for example, "Cauvery Water Row: Social Media Turns Anti-social," *Deccan Chronicle*, September 13, 2016, www.deccanchronicle.com/amp/nation/current-affairs/130916/cauvery-water-row-social-media-turns-anti-social.html.

5 See, for example, Sowmya Aji, "Karnataka Farmers in Distress as Crops in Karnataka Wither," *Economic Times*, October 7, 2016, https://economictimes.indiatimes.com/news/politics-and-nation/karnataka-farmers-in-distress-as-crops-in-cauvery-basin-wither/articleshow/54725081.cms. Tamil Nadu meanwhile has seen a similar acute crisis. During my fieldwork, for instance, there was a pattern of distress-related farmer suicides. In 2020–21, there has been a mass sustained mobilization of farmers in Delhi.

6 Chennai does not rely on the Cauvery resources for its drinking water supply. However, there are parallel patterns of strains on water resources through economic growth and both planned and unplanned urbanization.

7 Nagesh Prabhu, "No Water from Crops in Cauvery Basin," *Hindu*, August 2, 2017, www.thehindu.com/news/national/karnataka/no-water-for-crops-in-cauvery-basin/article19410286.ece.

8 For details on the final award, see "Award of Cauvery Tribunal," Ministry of Jal Shakti Department of Water Resources, River Development and Ganha Rejuvenation, Government of India, accessed December 5, 2021, http://jalshakti-dowr.gov.in/acts-tribunals/current-inter-state-river-water-disputes-tribunals/cauvery-water-disputes.

9 The central government established the Cauvery Management Authority in 2018 and took more than three years to appoint a chairperson for the entity. See "Cauvery Management Authority Gets Full-Time Chairman," *Hindu*, September 28, 2021, www.thehindu.com/news/national/tamil-nadu/cauvery-water-management-authority-gets-full-time-chairman/article36712190.ece.

10 "PM Modi Busy with Karnataka Polls, Says Centre after SC's Cauvery Order," *Deccan Chronicle*, May 3, 2018, www.deccanchronicle.com/nation/current-affairs/030518/pm-modi-busy-with-karnataka-elections-centre-after-scs-cauvery-order.html.

11. For a useful detailed timeline of events, see T. Arvind, "Cauvery Issue: A Timeline," *Hindu*, February 16, 2018, www.thehindu.com/news/resources/cauvery-issue-timeline/article11264946.ece1.
12. "Cauvery Issue: Centre Seeks Modification of SC Order to Set up Water Management Board," Firstpost, October 3, 2016, www.firstpost.com/india/kaveri-issue-centre-seeks-modification-of-sc-order-to-set-up-water-management-board-3031918.html.
13. "Can't Be Asked to Set up Cauvery Board: Centre," *New Indian Express*, October 4, 2016, www.newindianexpress.com/states/karnataka/2016/oct/04/cant-be-asked-to-set-up-cauvery-board-centre-1524853.html.
14. Samanwaya Rautray, "Supreme Court All Set to Give Cauvery Board Final Shape Soon," *Economic Times*, May 18, 2018, https://economictimes.indiatimes.com/news/politics-and-nation/supreme-court-all-set-to-give-cauvery-board-final-shape-soon/articleshow/64214965.cms.
15. For the 2019 amendment, see "The Inter-state River Water Disputes (Amendment) Bill, 2019," PRS Legislative Research, accessed December 5, 2021, https://prsindia.org/billtrack/the-inter-state-river-water-disputes-amendment-bill-2019.
16. R. Iyer had criticized the amendments for not incorporating an appeals process. This very lack of appeals process has in effect brought the Supreme Court back into the process as foreshadowed by Iyer. See Iyer 2002.
17. Chief Minister Jayalalitha attempted to gain the intervention of then Congress prime minister Manmohan Singh. T. Arvind, "Cauvery Issue: A Timeline," *Hindu*, February 16, 2018, www.thehindu.com/news/resources/cauvery-issue-timeline/article11264946.ece1; "Government Is Committed to Go Ahead with Mekedatu Dam: CM," *Business Standard*, August 8, 2015, www.business-standard.com/article/pti-stories/government-is-committed-to-go-ahead-with-mekedatu-dam-cm-115080800569_1.html.
18. "Cauvery Dispute: Is Resolution in Sight?," *Deccan Chronicle*, September 18, 2016, www.deccanchronicle.com/opinion/op-ed/180916/cauvery-dispute-is-resolution-in-sight.html.
19. The newly independent state of Telangana is now a part of this process.
20. L. Renganathan, "Troubled Waters of Cauvery: A Primer on the Legal Route," *Hindu*, October 31, 2015, www.thehindu.com/news/national/tamil-nadu/troubled-waters-of-cauvery-a-primer-on-the-legal-route/article7824627.ece.
21. Renganathan, "Troubled Waters of Cauvery."
22. This would eventually begin to change the dynamic of the Liaison Committee (Mohanakrishnan 2016b, 70).
23. "A Dream Will Come True Tomorrow," *Hindu*, September 28, 1996, cited in Mohanakrishnan 2011a, 144.
24. The initial goal of fifteen TMC was modified to twelve TMC.
25. "Andhra Pradesh in Dilemma over Tamil Nadu's Plea for Krishna Water," *Times of India*, January 19, 2015, http://timesofindia.indiatimes.com/india/Andhra-Pradesh-in-dilemma-over-Tamil-Nadus-plea-for-Krishna-water/articleshow/46049396.cms.

26 K. Raju, "Even Minor Maintenance Work Can't Be Done in Mullaperiyar Dam," *Hindu*, March 14, 2014, www.thehindu.com/news/national/tamil-nadu/even-minor-maintenance-works-cant-be-done-in-mullaperiyar-dam/article5781820.ece.

27 V. Mayilvaganani, "Madurai Corporation to Tap Water from Mullaperiyar Dam," *Times of India*, August 3, 2014, https://timesofindia.indiatimes.com/city/chennai/madurai-corporation-to-tap-water-from-mullaperiyar-dam/articleshow/39519366.cms.

28 "Two Years after SC Judgment, Mullaperiyar Dispute Drags On," *Economic Times*, August 13, 2016, https://energy.economictimes.indiatimes.com/news/renewable/two-years-after-sc-judgment-mullaperiyar-dispute-drags-on/53668129.

29 The most well known, of course, is the Narmada Bachao Andolan in Gujarat.

30 "Farmers' Protest Turns Violent," *Hindu*, December 22, 2011, www.thehindu.com/news/national/tamil-nadu/farmers-protest-turns-violent/article2736024.ece.

31 "Kerala CM's Dam Remark Draws Flak," *Hindu*, May 30, 2016, www.thehindu.com/news/national/kerala/Kerala-CM's-dam-remarks-draw-flak/article14347669.ece.

32 Via Onmanoraman, "Consider Reducing Mullaperiyar Water Level to 139 Feet: SC to Tamil Nadu," *Week*, August 16, 2018, www.theweek.in/news/india/2018/08/16/water-rising-mullaperiyar-states-supreme-court.html.

33 J. Venkatesan, "Kerala Blames Floods on TN's Sudden Water Releases," *Deccan Chronicle*, August 24, 2018, www.deccanchronicle.com/nation/current-affairs/240818/kerala-govt-blames-floods-on-tns-sudden-water-release.html.

34 J. Venkatesan, "Consider Reducing Water Level of Mullaperiyar Dam to 139 Ft," *Deccan Chronicle*, August 18, 2018, www.deccanchronicle.com/nation/current-affairs/180818/consider-reducing-water-level-of-mullaiperiyar-dam-to-139-ft.html.

35 "Kerala-TN Reach Accord over Parambikulam-Aliyar Waters," *New Indian Express*, October 16, 2012, www.newindianexpress.com/states/tamil-nadu/2012/oct/17/kerala-tn-reach-accord-on-parambikulam-aliyar-water-416083.html.

36 This has occurred in both 2013 and 2017, before and after the final Supreme Court judgment on the dam. See "Kerala Water Pact to Help Parched Kovai," *New Indian Express*, April 29, 2013, www.newindianexpress.com/states/tamil-nadu/2013/apr/29/kerala-water-pact-to-help-parched-kovai-472490.html.

37 For instance, a revenue official of Kerala removed a reading scale installed by TNPWD. "Kerala Official Removes Scale from Dam Site," *New Indian Express*, November 18, 2014; "Mullaperiyar Dam TNPWD Engineers Roughed up at Dam Site," *New Indian Express*, November 18, 2014; "Rumours Being Spread on Mullaperiyar Dam: PWD," *Hindu*, December 11, 2011; "Locals Stop PWD Team from Visiting Dam Site," *New Indian Express*, June 21, 2012.

38 "KERI [Kerala Engineering Research Institute] Expert Team Conducted a Survey in Mullaperiyar Reservoir Accompanied by Water Resources Dept That Produced an Objection by TNPWD," *Hindu*, September 15, 2011.

39 "TN Spy Found in Water Dispute with Kerala, Say Cops," Indo Asian news service, April 26, 2013.

40 "Kerala to Open Police Station Near Mullaperiyar Dam," *Hindu,* January 5, 2016, www.thehindu.com/news/national/tamil-nadu/Kerala-to-open-police-station-near-Mullaperiyar-dam/article13982607.ece.

41 "Tamil Nadu Withdraws Plea for CISF Security Cover at Mullaperiyar Dam," NDTV, April 13, 2016, www.ndtv.com/tamil-nadu-news/tamil-nadu-withdraws-plea-for-cisf-security-cover-at-mullaperiyar-dam-1395101.

4. EXTRACTION, INEQUALITY, AND WATER BUREAUCRACY

1 See, for example, "Delay in Opening Sluice Gates Caused Flooding," *Times of India,* December 9, 2015, https://timesofindia.indiatimes.com/india/delay-in-opening-sluice-gates-caused-flooding/articleshow/50099873.cms.

2 This is one of the stated objectives of the Water Resources Organisation (see PWD, n.d.a.).

3 PWD Demand no. 38 Policy Note 2005–6, 2006, 45.

4 At the time of publication, Census 2021 data was not available, and the census was reported to have been delayed by the COVID-19 pandemic.

5 One professor of civil engineering in Chennai noted that bureaucrats were resistance to new ideas, such as using porous materials rather than concrete for sidewalks. Author's interview, August 2017.

6 The Tamil Nadu government order was passed on November 11, 1991, and planning authorities require the permission of the Agricultural Department for the conversion of land. The order was also meant to protect wetlands.

7 Times News Network, "Comprehensive Infra Plan Needed in Suburbs, Say Experts," *Times of India,* December 10, 2012, https://timesofindia.indiatimes.com/city/chennai/comprehensive-infra-plan-needed-in-suburbs-say-experts/articleshow/17550328.cms.

8 "Level of Urbanisation," Ministry of Housing and Urban Affairs, Government of India, accessed December 7, 2021, http://mohua.gov.in/cms/level-of-urbanisation.php.

9 Karen Coelho has illustrated the ways in which this conception of public by Metrowater engineers encodes exclusionary conceptions of class-based respectability. See Coelho 2005b.

10 Census of India, Government of India, accessed December 5, 2021, www.census2011.co.in/census/metropolitan/435-chennai.html. The 2021 census was delayed.

11 C. S Kotteswaran, "Water from 300 Tiruvallur Wells for Parched City," *Deccan Chronicle,* February 5, 2017, www.deccanchronicle.com/nation/in-other-news/050217/water-from-300-tiruvallur-wells-for-parched-chennai.html.

12 Tamil Nadu Groundwater (Development and Management) Act, 2003, Casemine, accessed December 7, 2021, www.casemine.com/act/in/5a9ccdf34a932653478130aa.

13 T. Ramakrishnan, "Adieu to Tamil Nadu Ground Water Law," *Hindu,* September 20, 2013, www.thehindu.com/news/national/tamil-nadu/adieu-to-tamil-nadu-groundwater-law/article5147072.ece.

14 For a critical discussion of the central government's data on groundwater, see Dhawan 1995.
15 "Groundwater Declining in All the Districts of Tamil Nadu," *Deccan Chronicle*, January 3 2017, www.deccanchronicle.com/nation/current-affairs/030117/ground water-level-declining-in-all-the-districts-of-tamil-nadu.html.
16 National data on groundwater is also being tracked by NASA. See "Groundwater Gains in India," NASA Earth Observatory, accessed December 7, 2021, https://earthobservatory.nasa.gov/images/91008/groundwater-gains-in-india.
17 T. Ramakrishnan, "Adieu to Tamil Nadu Ground Water Law," *Hindu*, September 20, 2013, www.thehindu.com/news/national/tamil-nadu/adieu-to-tamil-nadu-groundwater-law/article5147072.ece.
18 "Tamil Nadu Brings Stringent Rules to Protect Groundwater," *Deccan Herald*, July 31, 2014, www.deccanherald.com/content/422966/tamil-nadu-brings-strin gent-rules.html; "TN Order Banning Water Extraction Challenged in High Court," *Business Standard*, October 23, 2014, www.business-standard.com/article/pti-stories /tn-order-banning-water-extraction-challenged-in-hc-114102300533_1.html.
19 "Indiscriminate Exploitation of Groundwater in TN: CAG Report," *Hindu*, August 13, 2014, www.thehindubusinessline.com/news/indiscriminate-exploi tation-of-ground-water-in-tn-cag-report/article23153804.ece.
20 "TN Order Banning Water."
21 For a detailed critical discussion of this process, see V. Sridhar, "A Pipe Dream?," *Frontline*, May 21, 2004, https://frontline.thehindu.com/other/article30222555.ece.
22 See "Sector-Wise GDP of India," Statistics Times, accessed August 25, 2017, http://statisticstimes.com/economy/sectorwise-gdp-contribution-of-india.php, for an overview.
23 This dominance is intensified both by the sociocultural power of larger cities and by the financial weakness of small urban areas (P. Mohanty 2016). Contestations within urban environments and between rural and urban contexts stand to deepen long-standing inequalities that structure access to water. David McKenzie and Isha Ray (2009, 443), for example, have demonstrated a national relationship between asset wealth and access to water. In Rajasthan, in one case, Rajputs in the village forced Dalits to pay, while in other villages, Dalits were excluded (see Cullet 2009, 167). Philippe Cullet also provides a useful analysis of the complex impact in terms of class inequality where the poorest are indirectly affected by cost-recovery principles.
24 Quoted in Amit Anand Choudhary, "Tamil Nadu Government Cannot Remain Silent on Farmers' Suicide: Supreme Court," *Times of India*, April 13, 2017, http://timesofindia.indiatimes.com/india/tamil-nadu-government-can-not-remain -silent-on-farmers-suicide-supreme-court/articleshow/58169114.cms.
25 See also Carolin Arul's research, which confirms this practice, noting that the state is able to "withdraw slowly the irrigation supply to the command areas and allow urbanization to take place on its own, so that agriculture becomes defunct" (2008, 197).

26 See, for example, "World Bank Approves $400 Million to Improve Urban Services in Tamil Nadu, India," World Bank Press Release, March 31, 2015, www.worldbank.org/en/news/press-release/2015/03/31/world-bank-improve-urban-services-tamil-nadu-india, and "World Bank Approves $150 Million Loan for Tamil Nadu," *Economic Times*, October 1, 2021.

27 K. Lakshmi, "Metrowater Plans Digital Water Meters in Commercial Buildings," *Hindu*, April 5, 2017, www.thehindu.com/news/national/tamil-nadu/metrowater-plans-digital-water-meters-in-commercial-buildings/article17819571.ece.

28 "All Commercial Establishments to Have Digital Water Meters Soon," *Hindu*, January 28, 2021, www.thehindu.com/news/cities/chennai/all-commercial-establishments-to-have-digital-water-meters-soon/article33679790.ece.

29 For an extensive critical discussion of this model of urban financing, see Kundu 2001.

30 This funding structure in Tamil Nadu was used as a national model in India and was specifically designed to address the lack of government resources necessary for infrastructure development in small towns.

31 Ashwin Mahalingam, Ganesh Devkar, and Satyanarayana Kalidindi (2011) have analyzed the microdynamics of such initiatives and have shown that there are varying degrees of effectiveness in the implementation of water-related infrastructure projects (including delays in the ability of local governments to effectively take over and manage completed projects).

32 These reforms were first implemented in TN in 1994 (GTN 1994). The Tamil Nadu Farmers' Management of Irrigation Systems Act, 2000 (TN Act 7 of 2001) was enacted on March 2001.

33 *World Bank Operations Evaluation Department Report*, ICRR 12141, July 20, 2005, http://documents.worldbank.org/curated/en/604181474649821377/pdf/000020051-20140613052041.pdf.

34 A. Vidyanathan, "Agrarian Crisis: Nature, Causes and Remedies," *Hindu*, November 8, 2006, www.thehindu.com/todays-paper/tp-opinion/agrarian-crisis-nature-causes-and-remedies/article3044873.ece.

35 For example, the city experienced significant floods in 1943, 1976, and 1985.

5. STATE, CLASS, AND THE AGENCY OF BUREAUCRATS

1 See, for example, Anna Hazare's movement, which gained widespread publicity in 2011. This highly publicized movement did not focus on corporate corruption.

2 Cited in Gopalkrishna Gandhi, "Governments Should Use Right to Transfer Officials Judiciously," *Hindustan Times*, February 22, 2014, www.hindustantimes.com/columns/governments-should-use-right-to-transfer-officials-judiciously/story-20t9ALideHXdNq3Ro2GMeL.html.

3 For an overview of such contemporary debates, see Vaishnav and Khosla 2016.

4 C. P. Chandrasekar and Jayati Ghosh, "The Withering Trend of Public Employment in India," *Hindu*, July 29, 2019, www.thehindubusinessline.com/opinion/columns/the-withering-tren-of-public-employment-in-india/article28750003.ece.

5   Government employees continue to resist this. See, for example, a 2017 strike by the Tamil Nadu Government Employees Association. The primary demand was a rollback of the new pension scheme. See "Government Employees Call Off Strike," *Hindu*, April 27, 2017, www.thehindu.com/todays-paper/tp-national/tp-tamilnadu/government-staff-call-off-strike/article18235867.ece.
6   Note that Modi in fact first used this term in the 2012 Assembly elections in Gujarat. See "Gujarat BJP Manifesto Targets Neo-middle class," *Deccan Herald*, December 3, 2012, www.deccanherald.com/content/296228/gujarat-bjp-manifesto-targets-neo.html.
7   Note that recent financial constraints have led the state government to try to cut back on such expenditures.
8   B. Sivakumar, "81 Lakh Tamil Nadu People in Search of Government Employment," *Times of India*, April 16, 2017, https://timesofindia.indiatimes.com/city/chennai/81-lakh-tamil-nadu-people-in-search-of-govt-employment/articleshow/58202411.cms.
9   Lisa Björkman's (2015) study provides useful insights into some of the challenges that technical workers face in Mumbai's water sector.
10 "Madras High Court Fumes over Unchecked, Corruption in Chennai Corporation," Legal India, August 16, 2018, www.legalindia.com/madras-high-court-fumes-over-unchecked-corruption-in-chennai-corporation.
11 "Desilting on Paper, Two Suspended," July 11, 2018, *Times of India*, https://timesofindia.indiatimes.com/city/delhi/desilting-on-paper-two-suspended/articleshow/64938021.cms.
12 Letter to editor, *Deccan Chronicle*, June 20, 2017, print edition.
13 Interviews were conducted with varying ranks of employees in January 2018.
14 The TNPWD Association of Engineers urged the state government to drop its plan to bifurcate PWD. See "Engineers against Bifurcation of Public Works Department," *Hindu*, October 25, 2009. The bifurcation was tied to a World Bank loan of Rs. 2,500 crores. See "Government Urged to Drop Move to Split PWD," *Hindu*, February 26, 2007.
15 There is a reservation of 30 percent for women in government jobs, provided for under section 21 of the Tamil Nadu State and Subordinate Service Rules.
16 The PWD does not publish data on its workforce. Estimates are based on interviews and field site discussions that I carried out in 2017.
17 Suresh Kumar, "Madras High Courts Quashes Appointment of 11 TNPSC Members," *Hindu*, December 23, 2016, www.thehindu.com/news/national/tamil-nadu/Madras-High-Court-quashes-appointment-of-11-TNPSC-members/article16927844.ece1.
18 For an exception, see A. Sinha's (2011) work on bureaucratic agency drawing on a rational actor model of choice.
19 See, for example, E. N. Mangat Rai (1973) and Mohammed Ali Rafath (2012). David Potter's (1996) work represents one of the few texts that has dealt in depth with similar memoirs, mainly in the colonial period and early decades of independence.

20 The book is not publicly for sale and is held within the offices of the PWD and within Anna University's Centre for Water Resources.
21 Medha Patkar is a well-known antidam activist who spearheaded a national and then international campaign against the dam on the Narmada River.

CONCLUSION

1 T. E. Narasimhan and Gireesh Babu, "Tamil Nadu Water Woes," *Business Standard*, June 17, 2019, www.business-standard.com/article/current-affairs/tamil-nadu-water-woes-chennai-goes-thirsty-industries-feel-the-heat-119061700389_1.html.
2 Hussain Zakeer, "Nature's Bounty: Mettur Dam Full after 5 Years," *Deccan Chronicle*, July 24, 2018, www.deccanchronicle.com/nation/current-affairs/240718/natures-bounty-mettur-dam-full-after-5-years.html.
3 "Stalin Slams AIADMK for Huge Wastage of Cauvery Waters," *Deccan Chronicle*, August 20, 2018, print edition.
4 Priyanka Thirumurthy, "Cauvery Delta Not Getting Enough Water from Mettur Dam: Farmers Slam TN Government," August 21, 2018, www.thenewsminute.com/article/cauvery-delta-not-getting-enough-water-mettur-dam-farmers-slam-tn-government-87021. For earlier, ongoing reports on problems with sand mining, see "Farmers Blame Water Crisis on 'Catastrophic' Sand Mining," *Times of India*, April 18, 2018, https://timesofindia.indiatimes.com/city/trichy/farmers-blame-water-crisis-on-catastrophic-sand-mining/articleshow/58230325.cms.
5 "Water Wars: Plachimada vs. Coca Cola," *Hindu*, July 15, 2017, www.thehindu.com/sci-tech/energy-and-environment/water-wars-plachimada-vs-coca-cola/article19284658.ece.
6 Sruthi Radhakrishnan, "The Hindu Explains: Sterlite Protests," *Hindu*, May 23, 2018, www.thehindu.com/news/national/tamil-nadu/the-hindu-explains-sterlite-protests/article23969542.ece; D. Govardan, "TN Government Orders Permanent Closure of Vedanta Group's Sterlite Plant in Tuticorin," *Times of India*, May 28, 2018, www.timesofindia.indiatimes.com/india/sterlite-plant-in-tuticorin-to-be-shut-down-permanently-orders-tn-govt/articleshow/64355727.cms.

WORKS CITED

PRIMARY SOURCES

British Colonial Government Documents

Bourne, John. 1856. *Public Works in India: Being a Letter to the Right Honourable Richard Vernon Smith, M.P.* London: Longman, Brown, Green, Longmans, and Roberts.

Cotton, Sir Arthur. 1854. *Public Works in India, Their Importance.* London: Richardson Brothers.

EIICC (East India Irrigation and Canal Company). 1860. *Prospectus—A Selection of Extracts from Various Official Reports and Documents, &c. Showing the Importance and Remunerative Character of Works of Irrigation and Water Transit in India.* United Kingdom: n.p.

GC (Governor in Council). 1856. *First Report of the Commissioners Appointed to Enquire into and Report upon the System of Superintending and Executing Public Works in the Madras Presidency, Submitted to the Governor in Council.* Madras: Church of Scotland Mission Press.

GM (Government of Madras). 1858. *Standing Orders of the Board of Revenue 1820-65 (Land Revenue Settlement and Miscellaneous).* Vol. 2. Government of Madras. September 23.

Grant, William. 1857. *The Circular and Standing Orders of the Department of Public Works from 1833 to 8th July 1857.* Madras: Press of the Society for Promotion of Christian Knowledge.

HD (Home Department). 1905. Summary of the *Lord Curzon of Kedleston, Viceroy and Governor General of India, in the Home Department. I.—January 1899 to April 1904. II.—December 1904 to November 1905.* Simla: Government Central Branch Press.

Maskell, John. 1856. "The Circular Orders of the Board of Revenue Issued during the Year 1856." Madras Presidency.

MP (Madras Presidency). 1852. *First Report of the Commissioner.* Madras Presidency: Government Press.

———. 1902. *Preliminary Report on the Investigation of Protective Irrigation Works and on Irrigation under Wells in the Madras Presidency.* Madras: Government Press.

———. 1907. *Report of the Administration of the Madras Presidency during the Year 1906–1907.* Madras: Government Press.

PWD. (Public Works Department) 1868. Summary of the Principal Measures Carried out in the Public Works Department during the Administration of Sir John L. M. Lawrence Viceroy and Governor General of India, from January 1864 to January 1869. Calcutta, India: Public Works Department Press.

———. 1869. *Civil Engineer Grievances in the Department.* London: Charing Cross.

———. 1898. *Summary of the Principal Measures of the ViceRoyalty of the Earl of Elgin in the Public Works,* 1898. Calcutta: Office of the Superintendent of Government Printing, India

———. 1903. *Report of the Indian Irrigation Commission.* London: Darling and Son.

PWDH (Public Works Department, Hyderabad). 1937. *Proceedings of Conference on the Proposed Tungabhadra Project.* PWD office Hyderabad. July 15–17.

PWDI (Public Works Department [Irrigation]). 1938. Government of Madras GO, no. 1372, June 27.

SB (Secretary of Bombay). 1935. PWD no. 7242-T, June 5, 1935, reproduced in Government of Madras Public Works Department (Irrigation) GO, no. 1327. June 27, 1938.

Tyrrell, L.-C. 1873. *Public Works Reform in India.* London: Edward, Bumpus, Holborn Bars.

Government of India Documents and Reports

ARC (Administrative Reforms Commission). 1973. *A Report on the Public Works Administration.* Vol. 1. Madras: Government of Tamil Nadu.

Census of India. 2011. "Chapter 1 Population, Size, and Decadal Change." Government of India. www.censusindia.gov.in/2011census/PCA/PCA_Highlights/pca_highlights_file/India/Chapter-1.pdf.

Census Organization of India. 2011. Census 2011 India. www.census2011.co.in.

Constituent Assembly of India. 1949. "Constituent Assembly Debates on 10 October, 1949 Part I." https://indiankanoon.org/doc/735670.

Government of India. 1949. Constitution of India. National Portal of India. www.india.gov.in/my-government/constitution-india/constitution-india-full-text.

———. 1988. "Executive Summary: India's Urbanisation Policy." *Habitat International* 12 (3): 99–102.

LS (Lok Sabha, Government of India). 2019. The Inter-state River Water Disputes (Amendment) Act, 2019. Lok Sabha, Government of India. http://164.100.47.4/Bills Texts/LSBillTexts/Asintroduced/46_2017_LS_Eng.pdf.

MF (Ministry of Finance). 2016. *Economic Survey 2015-16: Services Sector Remains the Key Driver of Economic Growth Contributing almost 66.1% in 2015-16*. Government of India. http://pib.nic.in/newsite/PrintRelease.aspx?relid=136868.

Ministry of Housing and Urban Affairs, Government of India. 2018. "Cities Profile of Round 1 Smart Cities." Smart Cities Mission. http://smartcities.gov.in/content/innerpage/cities-profile-of-20-smart-cities.php.

Ministry of Statistics and Programme Implementation Planning Commission, Government of India. 2017. "Sector-Wise Contribution of GDP of India" Statistictimes.com. http://statisticstimes.com/economy/sectorwise-gdp-contribution-of-india.php.

MP (Ministry of Personnel, Public Grievances and Pensions, & Department of Administrative Reforms & Public Grievances). 2010. "Civil Services Survey: A Report. Government of India." New Delhi: Department of Administrative Reforms & Public Grievances, Government of India. http://unpan1.un.org/intradoc/groups/public/documents/cgg/unpan046740.pdf.

MRD (Ministry of Rural Development, Government of India). 2002. "Guidelines on Swajaldhara, 2002." Department of Drinking Water Supply. http://ielrc.org/content/e0212.pdf.

MWR (Ministry of Water Resources). 1987. "National Water Policy, 1987." 2017. www.rainwaterharvesting.org/Downloads/nwp1987.pdf.

———. 2002. *Annual Report, 2001-2002*. http://jalshakti-dowr.gov.in/sites/default/files/anu41036322978_3.pdf.

———. 2002. "National Water Policy." New Delhi: Government of India. http://jalshakti-dowr.gov.in/sites/default/files/nwp20025617515534_1.pdf.

———. 2014. New Delhi. Government of India. *Annual Report 2013-2014*. http://mowr.gov.in/sites/default/files/AR_2013-14_1.pdf.

———. 2016. "Draft National Water Framework Bill, 2016." Ministry of Jal Shakti, Department of Water Resources, River Development, and Ganga Rejuvenation. http://mowr.gov.in/sites/default/files/Water_Framework_May_2016.pdf.

———. 2018. "History: Organizational History of the Ministry of Water Resources, River Development and Ganga Rejuvenation." Ministry of Jal Shakti, Department of Water Resources, River Development & Ganga Rejuvenation. http://jalshakti-dowr.gov.in/about-us/history.

Palat, Mohandas. 2003. "Swajladhara Guidelines." New Delhi: Ministry of Rural Development.

PC (Planning Commission Yojana Bhawan New Delhi). 2007. *Report of the Expert Group on "Ground Water Management and Ownership."* New Delhi: Government of India Planning Commission. www.indiawaterportal.org/sites/indiawaterportal.org/files/PC_Groundwater_0.pdf.

Prasad, Kamta. 2008. *Institutional Framework for Regulating Use of Ground Water in India*. Central Ground Water Board, Ministry of Water Resources, Government of India. http://cgwb.gov.in/INCGW/Kamta%20Prasad%20report.pdf.

PRI (Parliament of the Republic of India). 2002. *The Inter-state Water Disputes (Amendment) Act, 2002*. Parliament of the Republic of India. https://indiankanoon.org/doc/1048477.

Prime Minister's Office, Government of India. 2011. 2011 PM Speech Overview. Press Information Bureau. http://pib.gov.in/newsite/PrintRelease.aspx?relid=78562.

Reddy, D. Y. V. 2015. *The Fourteenth Finance Commission (FC-XIV)*. Government of India: Ministry of Housing and Urban Affairs.

SCI (Supreme Court of India). 2006. "In the Supreme Court of India Civil Original Jurisdiction Original Suit No. 3 of 2006." https://main.sci.gov.in/jonew/bosir/orderpdfold/1981163.pdf.

———. 2007. "Supreme Court of India Record of Proceedings Civil Appeal NO. 2453 /2007." Supreme Court of India. www.sci.gov.in/pdf/cir/2017-03-21_1490100979.pdf.

Sheelapriya, M. 2008. *Statistical Handbook of Tamil Nadu 2008*. Chennai: Principal Secretary and Director, Department of Economics and Statistics.

International Organizations Documents and Reports

Agrawal, Pronita Chakrabarti. 2008. *Performance Improvement Planning: Upgrading and Improving Urban Water Services*. Water and sanitation program overview paper. WSP. Washington, DC: World Bank. http://documents.worldbank.org/curated/en/2008/04/9520938/performance-improvement-planning-upgrading-improving-urban-water-services.

Baietti, Aldo, and Peter Raymond. 2005. *Financing Water Supply and Sanitation Investments: Utilizing Risk Mitigation Instruments to Bridge the Financing Gap*. Water Supply and Sanitation Sector Board discussion paper series, no. 4. Washington, DC: World Bank. http://documents.worldbank.org/curated/en/200/01/5730871/financing-water-supply-sanitation-investments-utilizing-risk-mitigation-instruments-bridge-financing-gap.

Baumann, Pari, Vasudha Chhotray, James J. Emery, John Malcolm Kerr, and Grant Milne. 2006. *Managing Watershed Externalities in India*. Washington, DC: World Bank. http://documents.worldbank.org/curated/en/2006/04/7047191/managing-watershed-externalities-india.

Burton, Martin, and Amarjit Sing Dhingra. 2014. *India: Support for the Implementation of the National Water Mission by State Governments in India*. Asian Development Bank. Vol. 1 (November) Main Report. New Delhi: Asian Development Bank.

CWR (Centre for Water Resources). 2003. *Role of Women, Small and Marginal Farmers as Stakeholders in Lower Bhavnani, Sathanur and Ponnaiyar Systems*. Institute of Water Studies, World Bank.

Foster, Vivien, Subhrendu Pattanayak, and Linda Stalker Prokopy. 2003. "Do Current Water Subsidies Reach the Poor?" *Water Tariffs & Subsidies in South Asia* (April 2003): 1–10. https://openknowledge.worldbank.org/bitstream/handle/10986/17260/349610CurrentowateroWaterTariffs.pdf?sequence=1&isAllowed=y.

Kacker, Suneetha Dasappa, Tracey Osborne Miller, and S. R. Ramanujam. 2014. *Running Water in India's Cities: A Review of Five Recent Public-Private Partnership Initiatives*. Washington, DC: World Bank Group. http://documents.worldbank.org/curated/en/2014/01/18882071/running-water-indias-cities-review-five-recent-public-private-partnership-initiatives.

Mcintosh, Arthur Charles, and Thelma A. Triche. 2009. *India—Improving Water Supply and Sanitation Services for the Urban Poor in India*. Washington, DC: World Bank. http://documents.worldbank.org/curated/en/2009/03/16704206/india-improving-water-supply-sanitation-services-urban-poor-india.

Morse, Bradford, and Thomas Berger. 1992. *Sardar Sarovar: The Report of the Independent Review*. International Environmental Law Research Centre. Ottawa: Resources Futures International.

OEDWB (Operations Evaluation Department, World Bank). 2005. *ICR Review: Report No. ICRR12141*, July 20, 2005.

Rajagopal, Srinivasan Raj. 2005. *Implementation Completion Report*, World Bank, Report No. 31721, May 31, 2005. Washington, DC: World Bank. https://documents1.worldbank.org/curated/en/112241468043455602/pdf/31721.pdf.

Swaroop, Ananda. 2011. *Trends in Private Sector Participation in the Indian Water Sector: A Critical Review*. Washington, DC: World Bank. http://documents.worldbank.org/curated/en/2011/01/15762923/trends-private-sector-participation-indian-water-sector-critical-review.

UNDP. 1985. *UNDP Report, Second Mission Report on Proposals for Planning and Management of Water Resources for Madras City 1085*. New York City: United Nations.

WB (World Bank). 1987. *India—National Water Management Project*. Washington, DC: World Bank. http://documents.worldbank.org/curated/en/1987/02/739735/india-national-water-management-project.

———. 1995a. *Staff Appraisal Report—India Tamil Nadu Water Resources Consolidation Project*. World Bank. http://documents.worldbank.org/curated/en/149291468750308623/pdf/multiopage.pdf.

———. 1995b. *Tamil Nadu Water Resources Consolidation Project*. World Bank. http://projects.worldbank.org/P010476/tamil-nadu-water-resources-consolidation-project?lang=en&tab=overview.

———. 1995c. *India—Second Madras Water Supply Project*. Washington, DC: World Bank. http://documents.worldbank.org/curated/en/1995/05/697332/india-second-madras-water-supply-project.

———. 1995d. *Learning from Narmada—Independent Evaluation Group (IEG)*. World Bank Group. http://documents.worldbank.org/curated/en/777211468249297544/pdf/28514.pdf.

———. 1997. *Implementation Completion Report-India—National Water Management Project*. Washington, DC: World Bank Group. http://documents.worldbank.org/curated/en/1997/01/732024/india-national-water-management-project.

———. 1998a. *India—Water Resources Management Sector Review: Groundwater Regulation and Management Report*. World Bank. http://documents.worldbank.

org/curated/en/372491468752788129/India-Water-resources-management-sector-review-groundwater-regulation-and-management-report.

———. 1998b. *India—Water Resources Management Sector Review: Report on the Irrigation Sector.* World Bank. http://documents.worldbank.org/curated/en/785521468775824994/India-Water-resources-management-sector-review-report-on-the-irrigation-sector.

———. 2000. *Politicians for Reform: Proceedings of the State Ministers Workshop on Rural Water Supply Policy Reforms in India.* Water and Sanitation Program–South Asia. New Delhi: World Bank.

———. 2001. *Capacity-Building for Sector Reforms: The Rural Program—India Country Team.* Washington, DC: World Bank. http://documents.worldbank.org/curated/en/2001/04/2171984/capacity-building-sector-reforms-rural-program-india-country-team.

———. 2004. *India- Second Madras Water Supply Project (English).* World Bank. http://documents.worldbank.org/curated/en/865341468750544187/India-Second-Madras-Water-Supply-Project.

———. 2005. *India—Tamil Nadu Water Resources Consolidation Project (English).* Washington, DC: World Bank. http://documents.worldbank.org/curated/en/112241468043455602/India-Tamil-Nadu-Water-Resources-Consolidation-Project.

———. 2014. *Running Water in India's Cities: A Review of Five Recent Public-Private Partnership Initiatives.* https://openknowledge.worldbank.org/bitstream/handle/10986/17747/843840WSP0Box30tnershipoInitiatives.pdf?sequence=1&isAllowed=y.

WBIEU (World Bank Infrastructure and Energy Unit). 2005. *Project Appraisal Document Third Tamil Nadu Urban Development Project.* https://documents1.worldbank.org/curated/en/521631468033574527/pdf/31839a.pdf.

World Water Council. 2018. "8th World Water Forum Roundtables Reports." World Water Council. www.worldwatercouncil.org/sites/default/files/Forum_docs/Ministerial_Roundtables_Report.pdf.

Tamil Nadu State Government Documents and Reports

Anbu, V. Irai. 2016. *Statistical Handbook of Tamil Nadu 2017.* Government of Tamil Nadu. August 1, 2018. www.tn.gov.in/deptst.

CE (Chief Engineer, State Ground Surface Water Resources Data Centre, WRO PWD). 2005. "Ground Water Perspectives: A Profile of Chennai District, 2005."

Chennai Metropolitan Water Supply & Sewerage Board. 2016. "Metro Water Expanded Chennai City Status of Water Supply Scheme." Chennai Metropolitan Water Supply & Sewerage Board. www.chennaimetrowater.tn.nic.in/pdf/NEW-CITY WSS.pdf.

EES (Employment Exchange Statistics–Live Register). 2021. Government of Tamil Nadu, March 4, 2021. www.tn.gov.in/sites/default/files/emp_statistics_Feb_2021.pdf.

GM (Government of Madras). 1958. *Standing Order of the Board Revenue (Land Revenue Settlement and Miscellaneous).* Vol 2. Madras: Government of Madras.

GTN (Government of Tamil Nadu). 1986. "Tamil Nadu Public Works Department Code." 3rd ed. Madras: Government of Tamil Nadu.

———. 1994. "Government of Tamil Nadu Order G.O. Ms no. (PED) 1184."

———. Expert Committee on Rainwater Harvesting and Safeguarding Groundwater. 2000. "Measures for Augmenting Drinking Water Resources for Chennai and Its Suburbs, Occasional Paper 1." Government of Tamil Nadu.

———. 2002. Tamil Nadu Government Gazette Extraordinary Act no. 37 of 2002. Government of Tamil Nadu. www.chennaimetrowater.tn.nic.in/pdf/eo219.pdf.

———. 2003. *Report of the Expert Committee on Development and Management of Water Resources of Tamil Nadu*. Vol. 1. Chennai: Water Resources Organization.

———. 2011. "Urban Scenario in Tamil Nadu Census of India 2011." www.tn.gov.in/cma/Urban-Report.pdf.

———. 2016. "Municipal Administration and Water Supply Department, Policy Note Demand no. 34, 2016-2017." http://cms.tn.gov.in/sites/default/files/documents/maws_e_pn_2016_17.pdf.

GWB (Government of Tamil Nadu Ground Water Branch). 1992. *Pallar River Basin: Water Users*. Institute for Water Studies, Madras.

IWS (Institute for Water Studies). 1994. "Agriculture Department, G.O. Ms no. 559, 09.10.91." *Water Resources Assessment and Management Strategies for Madras Basin*: A-26-A27. Madras: Government of Tamil Nadu.

Kafeel, Nilofer. 2016. *Labour and Employment Department Policy Note 2016-2017*. Chennai: Government of Tamil Nadu.

MMDA (Madras Metropolitan Development Authority). 1993. *Stormwater Drainage Masterplan for Madras City and the Feasibility Study for Madras Metropolitan Area Report*. Chennai: Government of Tamil Nadu.

Mohanakrishnan, A. 2011a. *History of the Krishna Water Supply Project for Chennai City Annexures and Maps*. Advisor to Government of Tamil Nadu (Water Resources).

———. 2011b. *History of the Krishna Water Supply Project for Chennai City*. Government of Tamil Nadu (Water Resources).

ORG (Operations Resource Group). 1994. *Rehabilitation Action Plan for Project Affected Persons of Mordhana Reservoir, Scheme*. Chennai: Public Works Department.

Palaniswami, E. K. 2017. "Public Works Department Irrigation Policy Note 2017-2018 Demand no. 40." Government of Tamil Nadu.

PWD (Public Works Department). n.d.a. "Water Resource Department—TNWRD." Accessed December 6, 2018. http://tnwrd.gov.in/water-resource-department.

———. n.d.b. "Institute for Water Studies Activities." Water Resources Department. www.wrd.tn.gov.in/IWS/IWS_activities.pdf.

———, Water Resources Organisation. 1994. "Water Resources Assessment and Management Strategies for Madras Basin." Institute for Water Studies, Madras. IWS Report no. 12/94.

———, Water Resources Organisation. 1997. "State Framework Water Resources Plan Annexure 1 Chennai Basin Group." Institute for Water Studies. IWS Report no. 3/97.

TNUDF (Tamil Nadu Urban Development Fund). n.d. "Tamil Nadu Urban Development Fund." September 6, 2016. www.tnudf.com/tnudf.asp.

———. 2016. *Annual Report 2015–2016*. Chennai: Tamil Nadu Urban Development Fund. http://www.tnuifsl.com/documents/annual_reports/ar1516.pdf.

———. 2020. *Annual Report 2019–2020*. Chennai: Tamil Nadu Urban Development Fund. http://tnuifsl.com/documents/annual_reports/ar1920.pdf.

TNUIFSL. n.d. "City Corporate Plan/City Development Plan/Town Investment Plan." http://tnuifsl.com/ccpbp.asp.

WRCP (Tamil Nadu Water Resources Consolidation Project, Management Consultancy and Technical Assistance). 2001. *Water Plan Palar River Basin Executive Summary Prepared by Tahel Consultancy Engineers and Associates, Assisted by IWS and WRO*. August 23.

WRD (Water Resources Department), PWD. 2018. "Participatory Irrigation Management—TNWRD." Government of Tamil Nadu. http://tnwrd.gov.in/web/operation-maintenance-website/participatory-irrigation-management.

WRO (Water Resources Organisation). 1996. *Technical Proposal for Consultancy Services Tamil Nadu Water Resources Consolidation Project*. Chennai: State of Tamil Nadu.

SECONDARY SOURCES

Agarwal, Anil, and Sunita Narain, eds. 1997. *Dying Wisdom: Rise, Fall and Potential of India's Traditional Water Harvesting Systems*. New Delhi: Centre for Science and Environment.

Agarwal, Bina. 1994. *A Field of One's Own: Gender and Land Rights in South Asia*. Cambridge: Cambridge University Press.

Aiyer, Ananthakrishnan. 2008. "The Allure of the Transnational: Notes on Some Aspects of the Political Economy of Water in India." *Cultural Anthropology* 22 (4): 640–58.

Akhmouch, Aziza. 2012. "Water Governance in Latin America and the Caribbean." *OECD Regional Development Working Papers*: 1–149.

Anand, Nikhil. 2017. *Hydraulic City: Water and the Infrastructures of Citizenship in Mumbai*. Durham, NC: Duke University Press.

Anbarasan, A. 2010. *Development of Operational Guidelines for Equitable Distribution of Surface Water to Chennai City*. Unpublished PhD thesis. Department of Civil Engineering, Anna University, Chennai.

Arul, Carolin. 2008. *Gaps in Irrigation Laws of Tamil Nadu*. Unpublished PhD thesis. Chennai: Center for Water Resources, Anna University.

Asthana, Vandana. 2009. *Water Policy Processes in India: Discourses of Power and Resistance*. New York: Routledge.

Asthana, Vandana, and A. C. Shukla. 2014. *Water Security in India: Hope, Despair, and the Challenges of Human Development*. New York: Bloomsbury.

Baer, Madeline. 2014. "Private Water, Public Good: Water Privatization and State Capacity in Chile." *Studies in Comparative International Development* 49 (2): 141–67.

Baietti, Aldo, and Peter Raymond. 2005. "Financing Water Supply and Sanitation Investments: Utilizing Risk Mitigation Instruments to Bridge the Financing Gap." *Water Supply and Sanitation Sector Board Discussion Paper Series* 4.

Bakker, Karen. 2010. *Privatizing Water: Governance Failure and the World's Urban Water Crisis*. Ithaca: Cornell University Press.

Ballabh, Vishwa, ed. 2008. *Governance of Water: Institutional Alternatives and Political Economy*. New Delhi: Sage.

Bandyopadhyay, Jayanta. 2016. "New Institutional Structure for Water Security in India." *Economic and Political Weekly* 51 (15): 15–17.

Bardhan, Pranab K. 1984. *The Political Economy of Development in India*. Oxford: Blackwell.

———. 2014. "Comparative Corruption in China and India." *Indian Growth and Development Review* 7 (1): 8–11.

Bauer, Carl. 2010. "Market Approaches to Water Allocation: Lessons from Latin America." *Journal of Contemporary Water Research & Education* 144 (1): 44–49.

Baviskar, Amita. 2004. "Between Micro-politics and Administrative Imperatives: Decentralisation and the Watershed Mission in Madhya Pradesh, India." *European Journal of Development Research* 16 (1): 26–40.

———. 2005. *In the Belly of the River: Tribal Conflicts over Development in the Narmada River*. New Delhi: Oxford University Press.

———. 2007. "The Dream Machine: The Model Development Project and the Remaking of the State." In *Growth, Equity, Environment and Population: Economic and Sociological Perspectives*, edited by Kanchan Chopra and C. H. Hanumantha Rao, 287–310. New Delhi: Sage.

Bear, Laura, and Nayanika Mathur. 2015. "Remaking the Public Good: A New Anthropology of Bureaucracy Introduction." *Cambridge Journal of Anthropology* 33 (1): 18–34.

Benbabaali, Dalal. 2008. "Questioning the Role of the Indian Administrative Service in National Integration." *South Asia Multidisciplinary Academic Journal*.

Benjamin, N. 1971. "Cauvery Water Dispute." *Economic and Political Weekly* 6 (34): 1794–95.

Bhambhri, Chandra Prakash. 1971. *Bureaucracy and Politics in India*. Delhi: Vikas.

Bhide, Amita. 2017. "Directed Decentralisation: Analysing the Experience of Decentralisation via JNNURM in Maharashtra." In *Social Dynamics of the Urban: Studies from India*, edited by N. Jayaram, 81–98. New Delhi: Springer.

Biswas, Asit K., and Kris Hartley. 2017. "From Evidence to Policy in India's Groundwater Crisis." *Diplomat*. https://thediplomat.com/2017/07/from-evidence-to-policy-in-indias-groundwater-crisis.

Björkman, Lisa. 2015. *Pipe Politics, Contested Waters: Embedded Infrastructures of Millennial Mumbai*. Durham, NC: Duke University Press.

BJP (Bharatiya Janata Party). 2014. *Ek Bharat Shreshtha Bharat Sabka Saath Sabka Vikas Election Manifesto 2014*. Bharatiya Janata Party. www.bjp.org/images/pdf_2014/full_manifesto_english_07.04.2014.pdf.

Bourdieu, Pierre. 1994. "Rethinking the State: Genesis and Structure of the Bureaucratic Field." *Sociological Theory* 12 (1): 1–18.

Brenner, Neil. 2004. *New State Spaces: Urban Governance and the Rescaling of Statehood.* Oxford: Oxford University Press.

Briscoe, John, and R. P. S. Malik. 2006. *India's Water Economy: Bracing for a Turbulent Future.* New Delhi: Oxford University Press.

Budds, Jessica, and Gordon McGranahan. 2003. "Are the Debates on Water Privatization Missing the Point? Experiences from Africa, Asia and Latin America." *Environment and Urbanization* 15 (2): 87–114.

Bussell, Jennifer. 2012. *Corruption and Reform in India: Public Services in the Digital Age.* New York: Cambridge University Press.

Butterworth, John, Raphaèle Ducrot, Nicolas Faysse, and S. Janakarajan, eds. 2007. *Peri-urban Water Conflicts: Supporting Dialogue and Negotiation.* The Netherlands: IRC International Water and Sanitation Centre.

Celio, Mattia, Christopher A. Scott, and Mark Giordano. 2010. "Urban-Agricultural Water Appropriation: The Hyderabad, India Case." *Geographical Journal* 176 (1): 39–57.

Chandra, Kanchan. 2015. "The New Indian State." *Economic & Political Weekly* 50 (41): 46–58.

Chatterjee, Partha. 2006. *The Politics of the Governed: Reflections on Popular Politics in Most of the World.* New York: Columbia University Press.

Chinnasamy, Pennan, and Govindsamy Agoramoorthy. 2015. "Groundwater Storage and Depletion Trends in Tamil Nadu State, India." *Water Resources Management* 29 (7): 2139–52.

Chng, Nai Rui. 2008. "Privatization and Citizenship: Local Politics of Water in the Philippines." *Development* 51 (1): 42–48.

Chokkakula, Srinivas. 2014. "Inter-state Water Disputes: Perils and Prospects of Democratisation." *Economic and Political Weekly* 9 (49): 75–81.

Coelho, Karen. 2005a. "The Political Economy of Public Sector Water Utilities Reform." *Infochange Agenda* 3: 13–17.

———. 2005b. "Unstating the 'Public': An Ethnography of Reform in an Urban Water Utility in South India." *The Aid Effect: Giving and Governing in International Development:* 171–95.

———. 2010. "The Slow Road to the Private: A Case Study of Neoliberal Water Reforms in Chennai." In *Water Governance in Motion: Towards Socially and Environmentally Sustainable Water Laws,* edited by Phillipe Cullet, A. Goawlland-Gualtieri, R. Madhav, and U. Ramanathan, 1–20. Delhi: Cambridge University Press.

———. 2017. "The Canal and the City: An Urban-Ecological Lens on Chennai's Growth." In *India's Contemporary Urban Conundrum,* edited by Sujata Patel and Omita Goyal, 287–300. London: Routledge.

Coelho, Karen, and N. Raman. 2013. "From the Frying Pan to the Floodplain: Negotiating Land, Water, and Fire in Chennai's Development." In *Ecologies of Urbanism in India: Metropolitan Civility and Sustainability,* edited by Anne Rademacher and K. Sivaramakrishnan, 145–68. Hong Kong: Hong Kong University Press.

Corbridge, Stuart, Glyn Williams, Manoj Srivastava, and René Véron. 2005. *Seeing the State: Governance and Governmentality in India*. Vol. 10. Cambridge: Cambridge University Press.

Cullet, Philippe. 2009. *Water Law, Poverty, and Development: Water Sector Reforms in India*. Oxford: Oxford University Press.

Das, S. K. 2001. *Public Office, Private Interest: Bureaucracy and Corruption in India*. New Delhi: Oxford University Press.

Dasgupta, Simanti. 2015. *BITS of Belonging: Information Technology, Water, and Neoliberal Governance in India*. Philadelphia: Temple University Press.

de Freitas, Corin. 2015. "Old Chico's New Tricks: Neoliberalization and Water Sector Reform in Brazil's São Francisco River Basin." *Geoforum* 64: 292–303.

Deshpande, Rajeshwari, K. K. Kailash, and Louise Tillin. 2017. "States as Laboratories: The Politics of Social Welfare Policies in India." *India Review* 16 (1): 85–105.

Dhawan, B. D. 1995. "Magnitude of Groundwater Exploitation." *Economic and Political Weekly* 30 (14): 769–75.

D'Souza, Radha. 2002. "At the Confluence of Law and Geography: Contextualising Inter-state Water Disputes in India." *Geoforum* 33 (2): 255–69.

———. 2009. "Nation vs. Peoples: Inter-state Water Disputes in India's Supreme Court." In *Water and the Laws in India*, edited by Ramaswamy R Iyer, 58–96. New Delhi: Sage.

D'Souza, Rohan. 2006. *Drowned and Dammed: Colonial Capitalism and Flood Control in Eastern India*. New Delhi: Oxford University Press.

Dubash, Navroz K. 2002. *Tubewell Capitalism: Groundwater Development and Agrarian Change in Gujarat*. New Delhi: Oxford University Press.

Dubash, Navroz K., and Bronwen Morgan, eds. 2013. *The Rise of the Regulatory State of the South: Infrastructure and Development in Emerging Economies*. Oxford: Oxford University Press.

Evans, Peter, and Patrick Heller. 2015. "Human Development, State Transformation and the Politics of the Developmental State." In *The Oxford Handbook of Transformations of the State*, edited by Stephan Leibfried, Evelyne Huber, Matthew Lange, Jonah D. Levy, Frank Nullmeier, and John D. Stephens, 691–713. Oxford: Oxford University Press.

Fernandes, Leela. 2004. "The Politics of Forgetting: Class Politics, State Power and the Restructuring of Urban Space in India." *Urban Studies* 41 (12): 2415–30.

———. 2006. *India's New Middle Class: Democratic Politics in an Era of Economic Reform*. Minneapolis: University of Minnesota Press.

———, ed. 2018a. *Feminists Rethink the Neoliberal State*. NY: New York University Press.

———. 2018b. "Inter-state Water Disputes in South India." *Oxford Research Encyclopedia in Asian History*, 1–25. https://doi.org/10.1093/acrefore/9780190277727.013.191.

Fioretos, Orfeo, Tulia G. Falleti, and Adam Sheingate. 2016. "Historical Institutionalism in Political Science." In *The Oxford Handbook of Historical Institutionalism*, edited by Orfeo Fioretos, Tulia G. Falleti, and Adam Sheingate, 3–30. Oxford: Oxford University Press.

Frankel, Francine R. 2015. *India's Green Revolution: Economic Gains and Political Costs*. Princeton: Princeton University Press.

Ganguly, Sumit, and William R. Thompson. 2017. *Ascending India and Its State Capacity: Extraction, Violence, and Legitimacy*. New Haven: Yale University Press.

Ganguly-Scrase, Ruchira, and Timothy J. Scrase. 2009. *Globalisation and the Middle Classes in India: The Social and Cultural Impact of Neoliberal Reforms*. New York: Routledge.

Ghuman, Ranjit Singh, and Rajeev Sharma. 2018. *Emerging Water Insecurity in India: Lessons from an Agriculturally Advanced State*. New Castle upon Tyne: Cambridge Scholars.

Gilmartin, David. 1994. "Scientific Empire and Imperial Science: Colonialism and Irrigation Technology in the Indus Basin." *Journal of Asian Studies* 53 (4): 1127–49.

Goldman, Michael, and Devika Narayan. 2019. "Water Crisis through the Analytic of Urban Transformation: An Analysis of Bangalore's Hydrosocial Regimes." *Water International* 44 (2): 95–114.

Gopakumar, Govind. 2012. *Transforming Urban Water Supplies in India: The Role of Reform and Partnerships in Globalization*. New York: Routledge.

Goswami, Manu. 2004. *Producing India: From Colonial Economy to National Space*. Chicago: University of Chicago Press.

Gould, William. 2011. *Bureaucracy, Community and Influence in India: Society and the State, 1930s–1960s*. New York: Routledge.

Guhan, Sanjivi. 1993. *The Cauvery Dispute: Towards Conciliation*. Madras: Frontline.

Guerrero, Tatiana Acevedo, Kathryn Furlong, and Jeimy Arias. 2015. "Complicating Neoliberalization and Decentralization: The Non-linear Experience of Colombian Water Supply, 1909-2012." *International Journal of Water Resources Development* 32 (2): 172–88.

Gupta, Akhil. 1998. *Postcolonial Developments: Agriculture in the Making of Modern India*. Durham, NC: Duke University Press.

———. 2012. *Red Tape: Bureaucracy, Structural Violence, and Poverty in India*. Durham, NC: Duke University Press.

———. 2017. "Changing Forms of Corruption in India." *Modern Asian Studies* 51 (6): 1862–90.

Gupta, Akhil, and Kalyanakrishnan Sivaramakrishnan, eds. 2011. *The State in India after Liberalization: Interdisciplinary Perspectives*. New York: Routledge.

Haldea, Gajendra. 2011. *Infrastructure at Crossroads: The Challenges of Governance*. Oxford: Oxford University Press.

Hall, David, Emanuele Lobina, and Robin de la Motte. 2005. "Public Resistance to Privatisation in Water and Energy." *Development in Practice* 15 (3–4): 286–301.

Hall, Peter, and Michèle Lamont, eds. 2013. *Social Resilience in the Neoliberal Era*. Cambridge: Cambridge University Press.

Harris, Leila M., and Maria Cecilia Roa-García. 2013. "Recent Waves of Water Governance: Constitutional Reform and Resistance to Neoliberalization in Latin America (1990–2012)." *Geoforum* 50: 20–30.

Harriss, John and Andrew Wyatt. 2019. "Business and Politics: The Tamil Nadu Puzzle." In *Business and Politics in India*, edited by Christophe Jaffrelot, Atul Kohli, and Kanta Murali, 234–59. New York: Oxford University Press.

Harriss-White, Barbara. 2016. *Middle India and Urban-Rural Development: Four Decades of Change*. New Delhi: Springer.

Heller, Patrick, Partha Mukhopadhyay, and Michael Walton. 2019. "Cabal City: Urban Regimes and Accumulation without Development." In *Business and Politics in India*, edited by Christophe Jaffrelot, Atul Kohli, and Kanta Murali, 151–82. New York: Oxford University Press.

Herzfeld, Michael. 1992. *The Social Production of Indifference: Exploring the Symbolic Roots of Western Bureaucracy*. Chicago: University of Chicago Press.

Hoque, Sonia Ferdous. 2012. "Urban Water Sector Reforms in India: Financing Infrastructure Development through Market-Based Financing and Private-Public Partnerships." Lee Kuan Yew School of Public Policy Research Paper no. LKYSPP: 12-05.

International Environmental Law Research Centre. 2006. *Kerala Irrigation and Water Conservation Act, 2003 (As Amended in 2006)*. www.ielrc.org/content/e0303.pdf.

Ioris, Antonio Augusto Rossotto. 2012. "The Neoliberalization of Water in Lima, Peru." *Political Geography* 31 (5): 266–78.

Islar, Mine, and Chad Boda. 2014. "Political Ecology of Inter-basin Water Transfers in Turkish Water Governance." *Ecology and Society* 19 (4): 15.

Iyer, Lakshmi, and Anandi Mani. 2012. "Traveling Agents: Political Change and Bureaucratic Turnover in India." *Review of Economics and Statistics* 94 (3): 723–39.

Iyer, Ramaswamy R. 2002. "Inter-state Water Disputes Act 1956: Difficulties and Solutions." *Economic and Political Weekly* 37 (28): 2907–10.

———. 2003. "Cauvery Water Dispute: A Dialogue between Farmers." *Economic and Political Weekly* 38 (24): 2350–52.

———, ed. 2009. *Water and the Laws in India*. New Delhi: Sage.

———. 2010. "Resolving River Water Disputes in India: Reflections." In *River Water Sharing: Transboundary Conflict and Cooperation in India*, edited by N. Shantha Mohan, N. Sashikumar, and Sailen Routray, 66–80. New Delhi: Routledge.

———. 2012. "River Linking Project: A Disquieting Judgment." *Economic and Political Weekly* 47 (14): 33–40.

———. 2013. "Viewpoint: The Story of a Troubled Relationship." *Water Alternatives* 6 (2): 168–76.

———, ed. 2015. *Living Rivers, Dying Rivers*. New Delhi: Oxford University Press.

Jaffrelot, Christophe, Atul Kohli, and Kanta Murali, eds. 2019. *Business and Politics in India*. New York: Oxford University Press.

Jairath, Jasveen. 2008. "Misgovernance of Droughts in India." In *Governance of Water: Institutional Alternatives and Political Economy*, edited by Vishwa Ballabh, 36–58. New Delhi: Sage.

Janakarajan, S. 2004. "A Snake in the Grass! Unequal Power, Unequal Contracts and Unexplained Conflicts: Facilitating Negotiations over Water Conflicts in Peri-urban Catchments." Paper presented at *Conference on Market Development of Water & Waste Technologies Through Environmental Economics*. www.irc.nl/negowat.

———. 2010. "Negotiation through Social Dialogue: Insights from the Cauvery Dispute." In *River Water Sharing: Transboundary Conflict and Cooperation in India*, edited by N. Shantha Mohan, N. Sashikumar, and Sailen Routray, 140–55. New Delhi: Routledge.

Janakarajan, S., John Butterworth, Patrick Moriarty, and Charles Batchelor. 2007. "Strengthened City, Marginalised Peri-urban Villages: Stakeholder Dialogues for Inclusive Urbanisation in Chennai, India." In *Peri-urban Water Conflicts: Supporting Dialogue and Negotiation*, edited by John Butterworth, Raphaèle Ducrot, Nicolas Faysse, and S. Janakarajan, 51–74. The Netherlands: IRC International Water and Sanitation Centre.

Jenkins, Rob, ed. 2004. *Regional Reflections: Comparing Politics across India's States*. New Delhi: Oxford University Press.

Jha, Prem Shankar. 1980. *India: A Political Economy of Stagnation*. Bombay: Oxford University Press.

Joy, K. J., Biksham Gupta, Suhas Paranjape, Vinod Goude, and Shruti Vispute. 2008. *Water Conflicts in India: A Million Revolts in the Making*. New Delhi: Routledge.

Kale, Sunila S. 2014. *Electrifying India: Regional Political Economies of Development*. Stanford: Stanford University Press.

Kapur, Devesh, and Pratap Bhanu Mehta. 2007. *Public Institutions in India: Performance and Design*. New Delhi: Oxford University Press.

Kapur, Devesh, Pratap Bhanu Mehta, and Milan Vaishnav, eds. 2017. *Rethinking Public Institutions in India*. New Delhi: Oxford University Press.

Kapur, Devesh, and Ravi Ramamurti. 2002. "Privatization in India: The Imperatives and Consequences of Gradualism." *Center for Research on Economic Development and Policy Reform, Working Paper no. 142*. http://policydialogue.org/files/publications/Privitization_in_India_Kapur.pdf.

Kemerink, Jeltsje Sanne, Stephen Ngao Munyao, Klaas Schwartz, Rhodante Ahlers, and Pieter van der Zaag. 2016. "Why Infrastructure Still Matters: Unravelling Water Reform Processes in an Uneven Waterscape in Rural Kenya." *International Journal of the Commons* 10 (2): 1055–81.

Kennedy, Loraine. 2004. "The Political Determinants of Reform Packaging: Contrasting Response in Andhra Pradesh and Tamil Nadu." In *Regional Reflections: Comparing Politics across India's States*, edited by Rob Jenkins, 29–65. New Delhi: Oxford University Press.

Khagram, Sanjeev. 2004. *Dams and Development: Transnational Struggles for Water and Power*. Ithaca: Cornell University Press.

Koonan, Sujith, and Lovleen Bhullar. 2012. "Water Regulatory Authorities in India: The Way Forward?" *International Environmental Law Research Centre Policy Paper (2012-04)*. http://ielrc.org/content/p1204.pdf.

Krishnan, L. 2007. "Tamil Nadu Urban Development Fund: Public-Private Partnership in an Infrastructure Finance Intermediary." In *Financing Cities Fiscal Responsibility and Urban Infrastructure in Brazil, China, India, Poland and South Africa*, edited by George E. Peterson and Patricia Clarke Annez, 238–62. New Delhi: Sage.

Kumar, Anil. 2011. "Federalism Yes, What about Decentralisation? Some Aspects of Politics and Governance in Andhra Pradesh and Tamil Nadu." *Journal of Polity and Society* 4 (1): 28–50.

Kumar, V. Anil. 2009. "Federalism and Decentralisation in India: Andhra Pradesh and Tamil Nadu." *Institute for Social and Economic Change* (working paper), 1–32.

Kundu, Amitabh. 2001. "Institutional Innovations for Urban Infrastructural Development: The Indian Scenario." *Development in Practice* 11 (2/3): 174–89.

Laurie, Nina, and Liz Bondi, eds. 2005. *Working the Spaces of Neoliberalism: Activism, Professionalisation and Incorporation*. Malden, MA: Blackwell.

Lipsky, Michael. 2010. *Street-Level Bureaucracy: Dilemmas of the Individual in Public Service*. 30th ann. ed. New York: Russell Sage Foundation.

Lobo, Lancy, Mrutuyanjaya Sahu, and Jayesh Shah, eds. 2014. *Federalism in India: Towards a Fresh Balance of Power*. Jaipur: Rawat.

Ludden, David E. 1979. "Patronage and Irrigation in Tamil Nadu: A Long Term View." *Indian Economic and Social History Review* 16 (3): 347–65.

———. 1985. *Peasant History in South India*. Princeton: Princeton University Press.

———. 1992. "India's Development Regime." In *Colonialism and Culture*, edited by Nicholas B. Dirks, 247–87. Ann Arbor: University of Michigan Press.

Madhusoodhanan, C. G., and K. G. Sreeja. 2010. *The Mullaperiyar Conflict*. Bangalore: National Institute of Advanced Studies, Conflict Resolution Programme.

Mahalingam, Ashwin, Ganesh A. Devkar, and Satyanarayana N. Kalidindi. 2011. "A Comparative Analysis of Public-Private Partnership (PPP) Coordination Agencies in India: What Works and What Doesn't." *Public Works Management & Policy* 16 (4): 341–72.

Mangat Rai, E. N. 1973. *Commitment My Style: Career in the Civil Service*. Delhi: Vikas.

Manor, James. 2001. "Center-State Relations." In *The Success of India's Democracy*, edited by Atul Kholi, 78–102. Cambridge: Cambridge University Press.

———. 2004. "User Committees: A Potentially Damaging Second Wave of Decentralisation?" *European Journal of Development Research* 16 (1): 192–213.

———. 2016. "India's States: The Struggle to Govern." *Studies in Indian Politics* 4 (1): 8–21.

McDonald, David A., and Greg Ruiters, eds. 2004. *The Age of Commodity: Water Privatization in Southern Africa*. London: Earthscan.

McDonald, Robert I., Katherine Weber, Julie Padowski, Martina Florke, Christof Schneider, Pamela A. Green, Thomas Gleeson, Stephanie Eckman, Bernhard Lehner, Deborah Balk, Timothy Boucher, Gunther Grill, Mark Montgomery. 2014. "Water on an Urban Planet: Urbanization and the Reach of Urban Water Infrastructure." *Global Environmental Change* 27: 96–105.

McKenzie, David, and Isha Ray. 2009. "Urban Water Supply in India: Status, Reform Options and Possible Lessons." *Water Policy* 11 (4): 442–60.

Mehta, Lyla. 2005. *The Politics and Poetics of Water: Naturalising Scarcity in Western India*. New Delhi: Orient Longman.

Min, Brian. 2015. *Power and the Vote: Elections and Electricity in the Developing World*. New York: Cambridge University Press.

Misra, Bankey Bihari. 1977. *The Bureaucracy in India: An Historical Analysis of Development up to 1947*. Delhi: Oxford University Press.

Mitchell, Timothy. 2002. *Rule of Experts: Egypt, Techno-politics, Modernity*. Berkeley: University of California Press.

Moench, Marcus, Elisabeth Caspari, and Ajaya Dixit, eds. 1999. *Rethinking the Mosaic: Investigations into Local Water Management*. Kathmandu: Nepal Water Conservation Foundation.

Mohan, N. Shanta, Sailen Routray, and N. Sashikumar. 2010. *River Water Sharing: Transboundary Conflict and Cooperation in India*. New Delhi: Routledge.

Mohanakrishnan, A. 2016a. *An Autobiography of Prof A. Mohanakrishnan. Part 1*. Chennai: Irrigation Management Training Institute.

———. 2016b. *An Autobiography of Prof A. Mohanakrishnan. Part 2*. Chennai: Irrigation Management Training Institute.

Mohanty, B. B., ed. 2016. *Critical Perspectives on Agrarian Transition: India in the Global Debate*. New York: Routledge.

Mohanty, Prasanna K. 2016. *Financing Cities in India: Municipal Reforms, Fiscal Accountability and Urban Infrastructure*. New Delhi: Sage.

Mollinga, Peter P. 2003. *On the Waterfront: Water Distribution, Technology and Agrarian Change in a South Indian Canal Irrigation System*. Hyderabad: Orient Longman.

Mooij, Jos, ed. 2005. *The Politics of Economic Reforms in India*. New Delhi: Sage.

Moore, Scott M. 2018. *Subnational Hydropolitics: Conflict, Cooperation, and Institution-Building in Shared River Basins*. New York: Oxford University Press.

Morgan, Bronwen. 2011. *Water on Tap: Rights and Regulation in the Transnational Governance of Urban Water Services*. New York: Cambridge University Press.

Mosse, David. 1999. "Colonial and Contemporary Ideologies of 'Community Management': The Case of Tank Irrigation Development in South India." *Modern Asian Studies* 33 (2): 303–38.

———. 2003. *The Rule of Water: Statecraft, Ecology, and Collective Action in South India*. New Delhi: Oxford University Press.

Nagaraj, R. 2014. "Public Sector Employment: What Has Changed?" *Conference on Political Economy of Contemporary India, Indira Gandhi Institute of Development Research, Mumbai* 20: 1–17.

———. 2015 "Can the Public Sector Revive the Economy? Review of the Evidence and a Policy Suggestion." *Economic and Political Weekly* 50 (5): 41–46.

Nariman, Fali S., 2009. "Inter-state Water Disputes: A Nightmare!" In *Water and the Laws in India*, edited by Ramaswamy R. Iyer, 32–57. New Delhi: Sage.

Nayar, P. K. B. 1969. *Leadership Bureaucracy and Planning in India: A Sociological Study*. New Delhi: Associated Publishing House.

North, Douglass. 1990. *Institutions, Institutional Change and Economic Performance*. Cambridge: Cambridge University Press.

Packialakshmi, S. 2012. "A Study on Groundwater and Its Techno-socioeconomic Implications." PhD dissertation, Department of Civil Engineering, Anna University.

Packialakshmi, Shanmugam, N. K. Ambujam, and Prakash Nelliyat. 2011. "Groundwater Market and Its Implications on Water Resources and Agriculture in the Southern Peri-urban Interface, Chennai, India." *Journal of Environmental Development Sustainability* 13 (2): 423–38.

Padhiari, Hemant Kumar, and Vishwa Ballabh. 2008. "Inter-state Water Disputes and the Governance Challenge." In *Governance of Water: Institutional Alternatives and Political Economy*, edited by Vishwa Ballabh, 174–94. New Delhi: Sage.

Pandian, Anand. 2003. "Ode to an Engineer." In *Waterlines: The Penguin Book of River Stories*, edited by Amita Baviskar, 12–27. New Delhi: Penguin Books.

Pani, Narendar, 2010. "Boundaries of Transboundary Water Sharing." In *River Water Sharing: Transboundary Conflict and Cooperation in India*, edited by N. Shantha Mohan, Sailen Routray, and N. Sashikumar, 47–65. New Delhi: Routledge.

Potter, David C. 1996. *India's Political Administrators: From ICS to IAS*. New York: Oxford University Press.

Powell, Walter W., and Paul J. DiMaggio, eds. 1991. *The New Institutionalism in Organizational Analysis*. Vol. 17, 1–38. Chicago: University of Chicago Press.

Prabhu, Nagesh. 2017. *Reflective Shadows: Political Economy of the World Bank Lending to India*. Oxford: Oxford University Press.

Prakash, Gyan. 1999. *Another Reason: Science and the Imagination of Modern India*. Princeton: Princeton University Press.

Prasad, Awadesh. 1976. *A Portrait of Bureaucracy in India*. Patna: Associated Book Agency.

Prasad, G. K. 1974. *Bureaucracy in India: A Sociological Study*. New Delhi: Sterling.

Punjabi, Bharat, and Craig A. Johnson. 2018. "The Politics of Rural-Urban Water Conflict in India: Untapping the Power of Institutional Reform." *World Development* 120: 182–92.

Rafath, Mohammed Ali. 2012. *Bureaucracy and Politics: Growth of Service Jurisprudence in All India Services*. Jaipur: Rawat.

Raina, Rajeswari S. 2015. "Technological and Institutional Change: India's Development Trajectory in an Innovation Systems Framework." In *Emerging Economies: Food and Energy Security, and Technology and Innovation*, edited by Parthasarathi Shome and Pooja Sharma, 329–51. New Delhi: Springer.

Rajagopal, A., and S. Janakarajan. 2006. "State in Perplexity: The Politics of Water Rights and Irrigation System Turnover in Tamil Nadu." *Water Nepal* 12 (1/2): 115–42.

Rajendran, S., and N. Rajasekaran. 2011. "Political Economy and Local Area Development Scheme in Tamil Nadu." *International Journal of Research in Commerce, Economics and Management* 1 (6): 32–35.

Rakendran, S., and S. Ramaswamy. 2017. "Court Restrains Water Sale from Tamirabarani in Tamil Nadu." *Economic & Political Weekly* 52 (9).

Ramadevi, R., and V. Balaraju Nikku. 2008. "Telugu Ganga Project: Water Rights and Conflicts." In *Water Conflicts in India: A Million Revolts in the Making*, edited by K. J. Joy, Suhas Paranjape, Shruti Vispute, Biksham Gujja, and Vinod Goud, 383–87. New Delhi: Routledge.

Richards, Alan, and Nirvikar Singh. 2002. "Inter-state Water Disputes in India: Institutions and Policies." *International Journal of Water Resources Development* 18 (4): 611–25.

Rudolph, Lloyd I., and Susanne Hoeber Rudolph. 1987. *In Pursuit of Lakshmi: The Political Economy of the Indian State*. Chicago: University of Chicago Press.

———. 2001. "Iconisation of Chandrababu: Sharing Sovereignty in India's Federal Market Economy." *Economic and Political Weekly* 36 (18): 1541–52.

Ruet, Joël, Marie Gambiez, and Emilie Lacour. 2007. "Private Appropriation of Resource: Impact of Peri-urban Farmers Selling Water to Chennai Metropolitan Water Board." *Cities* 24 (2): 110–21.

Saldhana, Leo F., and Bhargavi S. Rao. 2015. "Karnataka: Cauvery in Death Throes." In *Living Rivers, Dying Rivers*, edited by Ramaswamy R. Iyer, 293–313. New Delhi: Oxford University Press.

Salman, Salman M. A. 2002. "Inter-state Water Disputes in India: An Analysis of the Settlement Process." *Water Policy* 4 (3): 223–37.

Sampathkumar, T. Johnson. 2005. "Telugu Ganga Project: An Act of Inter-state Cooperation." *Indian Journal of Political Science* 66 (4): 851–72.

Sangita, S. N. 2014. "Interface of Local and Higher Governments: Nation Building and Inclusive Development in India." In *Federalism in India: Towards a Fresh Balance of Power*, edited by Lancy Lobo, Mrutuyanjaya Sahu, and Jayesh Shah, 62–95. Jaipur: Rawat.

Saraldevi, J. 2013. "People's Attitudes towards Paying for Water and Sanitation." PhD Dissertation, Faculty of Science and Humanities, Anna University Chennai.

Saravanan, V., and P. Appasamy. 1999. "Historical Perspectives on Conflicts over Domestic and Industrial Supply in the Bhavani and Noyyal Basins, Tamil Nadu." In *Rethinking the Mosaic Investigations into Local Water Management*, edited by Marcus Moench, Elisabeth Caspari, and Ajaya Dixit, 161–90. Kathmandu: Nepal Water Conservation Foundation.

Sassen, Saskia. 2001. *The Global City: New York, London, Tokyo*. Princeton: Princeton University Press.

Schmidt, Vivien. 2008. "Discursive Institutionalism: The Explanatory Power of Ideas and Discourse." *Annual Review of Political Science* 11 (1): 303–26.

Schnitzler, Antina von. 2008. "Citizenship Prepaid: Water, Calculability, and Techno-politics in South Africa." *Journal of Southern African Studies* 34 (4): 899–917.

Settar, S. 2010. "Kaveri in Its Historical Setting." In *River Water Sharing: Transboundary Conflict and Cooperation in India*, edited by N. Shantha Mohan, Sailen Routray, N. Sashikumar, 99–107. London: Routledge.

Shah, Mihir. 2013. "Water: Towards a Paradigm Shift in the Twelfth Plan." *Economic and Political Weekly* 48 (3): 40–52.

———. 2016. *A 21st Century Institutional Architecture for India's Water Reforms*. Report submitted by the Committee on Restructuring the CWC and CGWB.

http://cgwb.gov.in/INTRA-CGWB/Circulars/Report_on_Restructuring_CWC_CGWB.pdf.

———. 2018. "Resistance to Reforms in Water Governance." *Economic & Political Weekly* 53 (6): 60–63.

Shah, Mihir, and Himanshu Kulkarni. 2015. "Urban Water Systems in India: Typologies and Hypotheses." *Economic & Political Weekly* 50 (30): 57–69.

Shah, Mihir, and P. S. Vijayshankar, eds. 2016. *Water: Growing Understanding, Emerging Perspectives*. Hyderabad: Orient Blackswan.

Shah, R. B. 1994. "Inter-state River Water Disputes: A Historical Review." *International Journal of Water Resources Development* 10 (2): 175–89.

Shah, Tushaar. 2005. "The New Institutional Economics of India's Water Policy." Presentation made at International Workshop on African Water Laws: Plural Legislative Frameworks for Rural Water Management in Africa.

Sharma, Chanchal Kumar, and Wilfried Swenden. 2017. "Continuity and Change in Contemporary Indian Federalism." *India Review* 16 (1): 1–13.

Shiva, Vandana. 2016. *Water Wars: Privatization, Pollution and Profit*. New York: Penguin Random House.

Singh, Balmiki Prasad. 2017. *The Twenty-First Century: Geopolitics, Democracy and Peace*. New York: Routledge.

Singh, Prerna. 2016. *How Solidarity Works for Welfare Subnationalism and Social Development in India*. New York: Cambridge University Press.

Singh, Satyajit. 2002. *Taming the Waters: The Political Economy of Large Dams in India*. New York: Oxford University Press.

———. 2007. "Water and Local Governments: Institutional Design, Politics & Implementation." In *Decentralization: Institutions and Politics in Rural India*, edited by Satyajit Singh and Pradeep K. Sharma. Delhi: Oxford University Press.

———. 2016. *The Local in Governance: Politics, Decentralization and Environment*. Delhi: Oxford University Press.

Singh, Satyajit, and Pradeep K. Sharma, eds. 2007. *Decentralization: Institutions and Politics in Rural India*. Delhi: Oxford University Press.

Sinha, Aseema. 2005. *The Regional Roots of Developmental Politics in India: A Divided Leviathan*. Bloomington: Indiana University Press.

———. 2011. "An Institutional Perspective on the Post-liberalization State in India." In *The State in India After Liberalization: Interdisciplinary Perspectives*, edited by Akhil Gupta and K. Sivaramakrishnan, 49–68. New York: Routledge.

Srivastava, I. C., and B. D. Joshi, eds. 2012. *Bureaucracy in Action*. Jaipur: Rawat.

Steinmo, Sven. 2008. "Historical Institutionalism." In *Approaches and Methodologies in the Social Sciences: A Pluralist Perspective*, edited by Donatella Della Porta and Michael Keating, 118–39. New York: Cambridge University Press.

Stoddart, Brian. 2011. *Land, Water, Language and Politics in Andhra: Regional Evolution in India Since 1850*. New Delhi: Routledge.

Suresh, S. 2021. "Intersectoral Competition for Water Between Users and Uses in Tamil Nadu-India." *Frontiers in Earth and Science* 9 (September): 1–13. DOI: 10.3389/feart.2021.663198.

Swain, Ashok. 1998. "Fight for the Last Drop: Inter-state River Disputes in India." *Contemporary South Asia* 2 (7): 167–80.

Swenden, Wilfried, and Rekha Saxena. 2017. "Rethinking Central Planning: A Federal Critique of the Planning Commission." *India Review* 16 (1): 42–65.

Swyngedouw, Erik. 2004. *Social Power and the Urbanization of Water: Flows of Power.* Oxford: Oxford University Press.

Tarlo, Emma. 2003. *Unsettling Memories: Narratives of the Emergency in Delhi.* Berkeley: University of California Press.

Thateyus, A. J., Delphin Prema Dhanaseeli, and P. Vanitha. 2013. "Inter-state Dispute over Water and Safety in India: The Mullaperiyar Dam, a Historical Perspective." *American Journal of Water Resources* 1 (2): 10–19.

Tiwari, Piyush, and Ranesh Nair. 2011. "Transforming Water Utilities." In *India Infrastructure Report,* edited by Infrastructure Development Finance Company, 240–59. New Delhi: Oxford University Press.

Upadhyay, Videh. 2002. "Water Management and Village Groups: Role of Law." *Economic and Political Weekly* (37) 49: 4907–12.

———. 2009. "The Ownership of Water in Indian Laws." In *Water and the Laws in India,* edited by Ramaswamy R. Iyer, 134–48. New Delhi: Sage.

Urs, Kshithij, and Richard Whittell. 2009. *Resisting Reform?: Water Profits and Democracy.* New Delhi: Sage.

Vaddiraju, Anil Kumar. 2014. "Whither Decentralization in India?: The Interesting Story of a Nobody's Child." In *Federalism in India: Towards a Fresh Balance of Power,* edited by Lancy Lobo, Mrutuyanjaya Sahu, and Jayesh Shah, 96–111. Jaipur: Rawat.

Vaidyanathan, Anu. 1994. "Performance of Indian Agriculture Since Independence." In *Agrarian Questions,* edited by Kaushik Basu, 18–74. Delhi: Oxford University Press.

Vaidyanathan, A., and Bharath Jairaj. 2009. "Legal Aspects of Water Resource Management." In *Water and the Laws in India,* edited by Ramaswamy Iyer, 3–16. New Delhi: Sage.

Vaishnav, Milan, and Saksham Khosla. 2016. "The Indian Administrative Service Meets Big Data." New York: Carnegie Endowment for International Peace. https://carnegieendowment.org/2016/09/01/indian-administrative-service-meets-big-data-pub-64457.

Vasavi, A. R. 1999. *Harbingers of Rain: Land and Life in South India.* New Delhi: Oxford University Press.

Venkatachalam, Pritha. 2005. *Innovative Approaches to Municipal Infrastructure Financing: A Case Study on Tamil Nadu.* Working paper from Development Studies Institute London School of Economics and Political Science Working Paper Series.

Wade, Robert. 1982. "The System of Administrative and Political Corruption: Canal Irrigation in South India." *Journal of Development Studies* 18 (3): 287–328.

Warghade, Sachin, and Subodh Wagle. 2011. "Water Sector Reforms: Implications on Empowerment and Equity." In *India Infrastructure Report Water: Policy and*

Performance for Sustainable Development, edited by Infrastructure Development Finance Company, 325–36. New Delhi: Oxford University Press.
Washbrook, D. A. 1976. *The Emergence of Provincial Politics: The Madras Presidency.* New York: Cambridge University Press.
Wedeen, Lisa. 2002. "Conceptualizing Culture: Possibilities for Political Science." *American Political Science Review* 96 (4): 713–28.
Yates, Julian S., and Leila M. Harris. 2018. "Hybrid Regulatory Landscapes: The Human Right to Water, Variegated Neoliberal Water Governance, and Policy Transfer in Cape Town, South Africa, and Accra, Ghana." *World Development* 110: 75–87.

INDEX

Aam Aadmi Party (AAP), 177
Administrative Reforms Commission, 65–66
agency of bureaucrats. *See* bureaucratic agency
agriculture, 17, 31, 60, 74, 144, 166, 171, 187, 239n2, 247n25; water disputes and, 101, 107, 139–40. *See also* droughts; famine; farmers; floods; irrigation; rural areas
AIADMK (All India Anna Dravida Munnetra Kazhagam), 17, 112, 168, 203, 211, 240n10
Andhra Pradesh, 76, 89, 91, 100*map*, 103–4; interstate water, 18, 99, 101–2, 121–22, 125
Anna University: Centre for Water Resources, 210; Mohanakrishnan teaching at, 213
Arul, Carolin, 156, 168

Baer, Madeline, 26
Bakker, Karen, 26
Bangalore Water Supply and Sewage Board, 110
Bengaluru (Bangalore), 107, 110, 115, 142

BJP (Bharatiya Janata Party), 17, 87–88, 112, 117, 201–2. *See also* Modi, Narendra
Bombay Presidency, 52–53
Bourne, John, 35
British colonial state, 103, 107, 132
bureaucratic agency, 97, 119–20, 124, 190, 205, 220–24, 232, 249n18; institutional reforms and, 4–10, 19; in Tamil Nadu's PWD, 21–22, 39–42, 49–50, 195
bureaucrats, 4, 22, 99, 119–20, 192–95, 198–99, 212, 236, 246n5; corruption and patronage, 64–67, 193, 199–200, 203, 212, 231; middle-class formation for, 64–65, 201–5; rethinking the figure of, 205–11. *See also* engineers; Indian Administrative Service (IAS); middle class

canals, 59–60, 109, 126*map*, 156
caste, 177, 184–87, 210, 240n10. *See also* inequality; landless laborers; marginalized groups; poor people
Cauvery (Kaveri) River, 50–51, 53, 193, 108*map*, 225–26

273

INDEX

Cauvery (Kaveri) River dispute, 101, 105–7, 109–11, 119; Cauvery Family Initiative and, 118; historical roots of, 103–6; interstate water, 18, 102, 111–13, 115–17, 120–21, 132; Jayalalitha's fast and, 97–98; spaces of bureaucratic agency and, 119–20; state institutional incapacity and, 114–15, 117–18; Supreme Court interventions, 113–15

Cauvery Technical Cell, Tamil Nadu, 23–24, 119

Cauvery Water Management Board, 113–14

Cauvery Water Tribunal (Cauvery Tribunal), 98, 102, 106–7, 112, 116–17

central government, 5, 13–14, 32–33, 56–57, 61, 82, 84, 92, 95, 199, 240n7 (intro.); Indian bureaucrats and, 22; interstate, 99, 104–5, 111, 197; national river-interlinking project and, 86–87; Supreme Court and, 113–14, 169–70; as TNUIFSL funding source, 179–80. *See also* British colonial state; local governments; state governments

Central Ground Water Board (CGWB), 8–9, 57–58, 63, 86, 162

centralization, 33, 41–42, 55–57, 61, 109, 123, 142, 182; decentralization and, 27–28, 69; farmers' participatory irrigation management and, 88–89; interstate water negotiations and, 140–41; JNNURM and, 87; mechanisms, privatization and, 14–15; reforms and, 4, 242n4; in Tamil Nadu, 89–95; water governance in India and, 27, 177–78; World Bank and, 24, 72–73, 81–82. *See also* economic liberalization

Central Water Commission (CWC): Dam Safety Organization in, 135; establishment, 58, 63; Mullaperiyar Dam inspections by, 133; regulation of surface water and groundwater by, 57–58; reorganization need recognized, 86; Shah's report on restructuring, 8–9

Centre for Water Resources, Anna University, 210

Chandra, Kanchan, 200–201

Chembarambakkam Reservoir, Chennai, 143, 163, 192

Chennai, 21, 23, 91, 102, 121–24, 128–29, 142, 155, 158*map*, 176, 189–91, 204, 243n6; droughts and, 101, 143, 144, 225; flooding in, 143–44; state power and privatization in, 174–78. *See also* Chennai's water bureaucracy reforms

Chennai Metropolitan Area Ground Water (Regulation) Act (1987), 93, 159, 242n15

Chennai Metropolitan Water Supply and Sewerage Board (CMWSSB). *See* Metrowater, Chennai

Chennai's water bureaucracy reforms, 145–54

Chile, water privatization in, 26, 241n2

cities, metropolitan, 88, 142–43, 163

citizenship rights: water governance study and, 12

civil service. *See* Indian Administrative Service (IAS)

civil society, 19, 54, 118, 136. *See also* protests and protest movements; social movements

class, 169, 185–87, 203, 208–10, 214–15, 247n23. *See also* caste; inequality; middle class

climate change, x, 5, 101, 129, 144, 190, 233, 236–37. *See also* environment

Coca-Cola, protests against, 174, 230–31

Coelho, Karen, 153, 246n9

Coimbatore, 60, 90, 138, 147–48

Colombia, water governance in, 27

Congress Party, 54, 94, 106, 117, 122, 197

constitution (India): on interstate water disputes, 56; on water governance, 13, 32, 55–56, 82, 89, 95

INDEX 275

corruption, 7, 21–22, 39, 63–65, 177, 193–94, 197, 200, 203, 206–7, 212, 231–32, 248n1
cost recovery, 77, 84, 86, 90, 176, 242n7. *See also* Metrowater, Chennai; revenue collection; user fees; water charges; water metering
Cotton, Arthur, 30–31, 35–37, 49
Cullet, Philippe, 241n8

Dalhousie, Governor General, 35
dams, 57, 105, 109, 115, 209–10, 214–16, 220–23; World Bank funding for, 73, 75, 77. *See also* large infrastructure projects; Mettur Dam; Mullaperiyar Dam; Narmada River dam; Sardar Sarovar Dam project; Siruvani Dam
decentralization, 3, 5, 13–14, 16, 21, 69–70, 80, 82–84, 95, 106, 140–42, 145, 150, 162–63, 181–82, 228, 239n3, 240n7 (intro.); in Tamil Nadu, 9–10, 179; Water Resources Consolidation Projects and, 147–48; World Bank policies and projects in India and, 24, 72, 77, 80–81. *See also* economic liberalization; privatization; Water Users' Associations
Delhi, 20, 142, 177
Delhi Jal (water) Board, 229–30
developmental state, 2, 13–14, 32, 54–63, 67–68, 101, 105, 197
Devkar, Ganesh, 248n31
DMK (Dravida Munnetra Kazhagam), 17, 123–24, 203, 240n10
drinking water, 58, 60, 82–85, 90, 102, 105, 110–11, 138–40, 144, 151–52, 165–66, 168, 171, 226. *See also* Telugu Ganga Project
droughts, 129, 143, 144, 155–56, 162, 190, 207, 225, 236, 241n18; climate change and, 129, 144, 236; farmer suicides and, 169–70; interstate, 101, 138, 245n36. *See also* Cauvery (Kaveri) River dispute; famine; Mullaperiyar Dam conflict; water scarcity
Dubash, Navroz, 11

economic growth, 2–3, 235, 239n2
economic liberalization, 111, 134, 142, 152, 190, 200–201, 227. *See also* postliberalization period
educational capital, 208–9, 215–16
electoral politics, 187–89, 201–2, 249n6. *See also* political parties
electricity, 105, 171–72. *See also* hydropower
Emergency Rule, 122–24, 196, 197
employment assistance program (MGNREGA), 2
employment exchanges, 204–5
engineers, 40–41, 43–45, 64–66, 120, 128, 164, 175–76, 192, 208–9, 246n9. *See also* bureaucrats; Iyer, Ramaswamy; Mohanakrishnan, A.; Pennycuick, John; professional expertise
environment, 11–12, 75, 82–83, 135–36, 139, 173, 245n29. *See also* climate change
ethics, 214, 220–22, 232–33
ethnic groups: linguistic reorganization of states and, 103–4; violence among, Cauvery River dispute and, 107
expertise: PWD's use of term, 43–47. *See also* professional expertise
extraction, state-based: bureaucratic authority redistribution, inequality and, 154–57, 159–66; groundwater, 58, 59–60, 67, 105, 166; planned economy and, 57; privatization in the water sector and, 15, 16

famine, 41–42, 48–49, 240n6. *See also* droughts; water scarcity
Famine Commissions, 48, 49
farmers: 64, 75–76, 83, 90, 106, 109–10, 116, 118, 123, 129, 136, 167, 169–72, 185–87, 243n5. *See also* agriculture; irrigation; landless laborers

federalism (India): 3, 7, 10, 13–14, 21, 55, 62, 71, 96–99, 101, 103, 112; World Bank on institutional reforms and, 78–79. *See also* constitution (India); interstate water disputes

floods, 137, 143–44, 148–49, 190, 192, 236, 241n18, 248n35. *See also* Cauvery (Kaveri) River dispute; Mullaperiyar Dam conflict

Gandhi, Indira, 122–23, 196, 197–98
Gandhi, Mahatma, 54
Gandhi, Rajiv, 82–83
gender, groundwater markets and, 186–87. *See also* women
Gopakumar, Govind, 164, 239n4
Gould, William, 64–65
Government of Tamil Nadu (GoTN), 148*fig.*, 152–53, 171, 178; Second Tamil Nadu Urban Development Project, 91–92; Water Resources Consolidation Project in, 90–91
Green Revolution, 32, 74
groundwater, 32, 56–60, 67, 87, 105, 155–57, 159–61, 163–65, 166, 171–73, 173*fig.*, 241n8; Easement Act (1882) on, 240n7 (ch. 1); legislation in Tamil Nadu, 93–94; NASA tracking of, 247n16; overextraction of, 8–9, 63
Guhan, S., 120
Gujarat: Narmada Bachao Andolan in, 245n29; Narmada River dam project in, 75

Haldea, Gajendra, 20–21
Hazare, Anna, 248n1
Herzfeld, Michael, 194
Hoque, Sonia, 179
Hyderabad, 51, 52–53, 103, 142, 241nn9–10
hydropower, 132, 134, 221. *See also* dams

Idduki Reservoir, 130*map*, 137
identity, bureaucratic agency and, 211–20

Indian Administrative Service (IAS), 18, 194, 195–200, 240n1. *See also* bureaucratic agency
Indian Civil Service (ICS), 196–97
industry, 4–5, 144, 151–52, 166, 167*table*
inequality, x, 4–10, 12, 24–28, 64, 88, 110–11, 153–57, 159–66, 168–69, 182–89, 225, 236–37; institutional and, 4, 145, 227, 228–29, 235; privatization and, 10, 15–16. *See also* caste; institutional reforms; marginalized groups
infrastructural security, Mullaperiyar Dam conflict and, 131–40, 245n37
infrastructure, water-related, 11–12, 60, 87, 180, 218. *See also* canals; dams; large infrastructure projects; public works; Tamil Nadu Urban Infrastructure Financial Services Limited
Institute for Water Studies, 205, 208, 209
institutional reforms, ix, 21, 70–71, 77, 83, 90–91, 145, 147, 189–90, 227–28, 230, 233–35; inequality, bureaucratic agency, and, 4–10; national water policy framework and, 83; new forms of centralization and, 146, 153–54; in small towns and rural communities, 178–82
integrated water management, governance policies and principles based on, 234–35
interstate cooperation, on Telugu Ganga Project, 121–25, 126*map*, 127–29, 131, 244n24
Inter-state River Water Disputes Act (1956), 56, 104, 105, 113–14; amendments (2002), 114, 244n16
interstate water disputes, 55–56, 63, 94, 99, 101–6, 234, 242n2. *See also* Cauvery (Kaveri) River dispute; Mullaperiyar Dam conflict
interstate water negotiations, 53–54, 96–99, 124–25, 128–29, 244n24
irrigation, 30–31, 38–39, 41, 43, 45–46, 48, 50, 54, 59–60, 62–64, 84, 105, 107, 109,

125, 132, 146, 157, 166–67, 167*table*, 196, 226, 240n4, 240n7 (ch. 1); Madras government, 33, 35; PWD, 20, 39, 61*table*, 146, 156–57; World Bank support for, 74. *See also* dams; droughts; famine
IT sector, 151, 169, 172–73, 239n2, 242n1
Iyer, Ramaswamy, 220–24, 244n16

Janakarajan, S., 118
Jawaharlal Nehru National Urban Renewal Mission (JNNURM), 87, 88
Jayalalitha, Chief Minister, 97, 98, 204, 244n17
Jha, Prem Shankar, 198

Kalidindi, Satyanarayana, 248n31
Kapur, Devesh, 18
Karnataka, 79–80, 99, 100*map*, 103–4, 106–7, 109, 112, 118, 122–23; interstate water-related agreements and, 101–2, 104, 122. *See also* Cauvery (Kaveri) River dispute
Kaveri River. *See* Cauvery (Kaveri) River
Kejriwal, Arvind, 241n1
Kerala, 18, 99, 104, 130*map*, 137. *See also* Mullaperiyar Dam conflict
Khosla, Saksham, 200
Krishnan, L., 180
Krishna Water Supply Project, 102, 213. *See also* Telugu Ganga Project
Krishna Water Tribunal, 122, 244n19

Lakkavalli Reservoir, Myosore, 51–52, 53
land, 49, 156–57, 162–63, 166–74, 183. *See also* wetlands development
Land Acquisition Act (1894), 60, 156
Land Acquisition and Economic Rehabilitation (LAER) Office, 183–85
landless laborers, 172, 183, 186
large infrastructure projects, 35, 87, 209–10. *See also* canals; dams; infrastructure, water-related; public works

Lawrence, John, 43
Liaison Committee, for Telugu Ganga Project, 124–25, 244n22
liberalization. *See* economic liberalization
linguistic groups, reorganization of states and, 103–4
Lloyd George, David, 196
lobbying, bureaucratic, 35, 37–38, 39, 42–43, 47
local governments, 5, 16, 18, 85–86, 239n3. *See also* central government; metropolitan cities; Panchayati Raj Institutions (PRI); *panchayats*; small towns; state governments
local state governments. *See* state governments
Lok Sabha, 114. *See also* Parliament

Madhya Pradesh, 50, 79–80, 89, 91, 240n7 (ch. 1)
Madras. *See* Chennai
Madras High Court, 59, 167–68, 170, 207, 211
Madras Metropolitan Corporation (MMC), 148–49
Madras Presidency, 17, 19, 31–33, 34*map*, 35–39, 51–54, 103, 107, 132, 241nn9–10. *See also* Andhra Pradesh
Madurai, 132–33, 147–48
Mahalingam, Ashwin, 248n31
Maharashtra, 79–80, 89, 91, 100*map*, 122
marginalized groups, 82–83, 176–77, 236. *See also* inequality
markets. *See* water markets
media, 135, 136, 212
Mehta, Pratap Bhanu, 18
metropolitan cities, 88, 142–43, 163
Metrowater, Chennai, 21, 24, 102, 146–47, 148*fig.*, 149, 154–56, 159–61, 163–64, 167, 172, 175–77, 181, 186, 189, 226, 229; groundwater regulation in Chennai area by, 160, 161–62, 163; public distrust of, 207, 231

Mettur Dam, 225–26
middle class, 64–66, 169, 195, 201–5, 215–16. *See also* bureaucrats
Ministry of Water Resources, 56, 62–63, 113, 220–22, 261n17
Modi, Narendra, 2, 86–87, 94, 112, 201–2, 249n6. *See also* BJP
Mohanakrishnan, A., 23–24, 98–99, 119, 120, 125, 127–28, 193, 212–19, 232, 250n20
Mollinga, Peter, 240n6
monsoons, 101, 113, 225–26. *See also* droughts
Mordhana Reservoir scheme, 184–85
Morgan, Bronwen, 11
Mosse, David, 38, 46
Mullaperiyar Dam, 102, 130*map*, 213
Mullaperiyar Dam conflict, 131–40, 226; revenue reading scale and, 245n37
Municipal Administration and Water Supply (MAWS), 148*fig.*
Mysore, 51–52, 103, 107

Nagaraj, R., 203
Nair, Ranesh, 87
Narmada Bachao Andolan, 75
Narmada River dam, 73, 75, 221, 222
National Water Policy (India), 62, 86, 87, 168
Nehru, Jawaharlal, 32, 57, 214, 216
neoliberalism, 25, 229, 233
New Veeranam Extension Project, 168
NGOs (nongovernmental organizations), 81, 118, 136

Orissa, 38, 77, 89, 100*map*, 240n3
Other Backward Classes and Scheduled Castes/Dalits (OBC/SC castes), 184

packaged drinking water industry, 165
Panchayati Raj Institutions (PRI), 83, 182
panchayats, 2, 84–85
Pandian, Anand, 223
Pani, Narendar, 106

Parliament, 13, 18, 113–14
participatory development, 78–79, 83. *See also* decentralization; Water Users' Associations
participatory management, 88–89, 92, 177–78, 181, 210. *See also* decentralization; Water Users' Associations
Patel, Sardar Vallabhai, 195–99, 240n1
Patkar, Medha, 221, 250n21
patronage, 15–16, 63–64, 66, 198–200, 217–18, 233
Pennycuick, John, 223
peri-urban areas, 143, 155–56, 160, 166–67, 172–73, 173*fig.*, 187
planning, centralized: interstate water disputes and, 105; River Boards Act (1956) and, 104; water governance and, 237
planning and development, 56, 61, 82, 104
police, Mullaperiyar Dam conflict and, 138–40
political economy, 31–32, 99, 105, 111–13, 117–18, 135, 141, 164
political parties, 99, 106–7, 112, 117–20, 197–98, 200, 218. *See also entries for specific parties*
political power, 142, 164–65, 240n2. *See also* state power
politics of water, 24–28, 196–97, 226–27, 233; modernist vision of developmental state and, 32, 57
Poondi Reservoir, Chennai, 157, 158*map*
poor people, 153–54, 177, 183, 188–89. *See also* caste; marginalized groups
population growth, 142, 148, 155, 246n4
populist politics, ix, 203–4
postliberalization period, 1–2, 5–6, 10–16, 21, 71, 82–83, 88, 94–95, 101, 105–6, 140, 145, 152, 169, 239n1; reforming the bureaucracy in, 199–201
post-Washington Consensus, ix, 25, 229
Potter, David, 249n19
Prabhu, Nagesh, 74, 78, 229–30

Prasad, A., 212
PRI (Panchayati Raj Institutions), 83, 182
princely states, colonial state disputes with, 52
private interests (sector), 10, 25–26, 36, 83, 88–92, 111, 200, 228–29. *See also* public-private partnerships (PPPs)
privatization, 4, 14–15, 24–25, 69, 80, 82, 176–77, 229–31, 241n1; in Chennai, 156, 174–78; Chile's expansion of, 241n2; state power in small towns and rural areas and, 178–82; in Tamil Nadu, 9–10; water governance and, 3, 5, 26–27; water reform debates on, 154–55; World Bank and, 24, 72–73, 78, 80–81. *See also* decentralization; public-private partnerships
professional expertise, 41–47, 53. *See also* engineers
project-affected persons (PAPs), land acquisition for water infrastructure and, 183
property rights, 50, 241n8. *See also* groundwater; land
protests and protest movements, 25, 54, 75, 81, 97–98, 168, 186, 188, 225, 250n21; over Cauvery River dispute, 106–7, 113, 117; against private companies, 174, 230–31. *See also* civil society; Iyer, Ramaswamy
public goods, 154, 155, 177, 187
public health, national water policy framework and, 82–83
public interest, 35–36, 47–52. *See also* public welfare
public-private partnerships (PPPs), 78–80, 88, 178–80, 241n3; for water-related projects, 70, 174–75. *See also* privatization
public sector, 216–17. *See also* bureaucrats
public welfare, 168–70. *See also* public interest

public works, 30–31, 33, 35–39, 54, 241n13; Tamil Nadu's water bureaucracy and, 17–22. *See also* canals; dams; infrastructure, water-related; large infrastructure projects
Public Works Department (PWD), x, 19–21, 23–24, 29–31, 37, 39–40, 59–60, 65–66, 117, 124, 139, 146–48, 148*fig.*, 153, 155, 163, 205, 207, 209–10, 229, 231, 241n13; expanding authority of, 47–55; internal hierarchies within, 207–8; irrigation budget 1967–1973, 61*table*; professionalism, expertise and institutional power of, 44–47; staff reductions, 175–76. *See also* Mohanakrishnan, A.

racialized institutional norms, for professional expertise, 45–47
Rama, N. T. (NTR), 121, 127
Ramachandran, M. G. (MGR), 121, 127
Raman, N., 153
Rao, Bhargavi, 110
Redhills Lake, Chennai, 157, 158*map*
regulatory state, shift from developmental state in postliberalization period to, 10, 11, 227–28
rehabilitation, economic, land acquisition for water infrastructure and, 75–76, 161, 183–85
rent-seeking, privatization in the water sector and, 15, 16
reservoirs, rain-fed, 155, 156, 209–10
revenue collection, 83, 176–77, 196. *See also* cost recovery; user fees; water charges; water metering
Revenue Department, 40–41, 59, 148*fig.*
revenue sources, 30–31, 33, 35, 36, 48–49
River Boards Act (1956), 56, 104, 109, 135
rivers, 17–18, 47, 94, 101, 146, 149–50; interlinking project, 14, 86–87, 242n9. *See also* Cauvery (Kaveri) River; Interstate River Water Disputes Act; Telugu Ganga Project

Rudolph, Lloyd and Susanne, 240n7 (intro.)
rural areas, 9, 16, 20, 23, 79, 83–84, 142, 144–45, 161, 169, 179, 183, 185, 206; private capital, reforms, and state power in, 178–82. *See also* poor people; small towns
Rural Water Supply Programme of Government of India, 84

Saldhana, Leo, 110
Sangita, S. N., 85
sanitation services, 71, 74–75, 76, 87. *See also* sewers
Sardar Sarovar Dam project, 75, 79, 221, 222
security, Mullaperiyar Dam conflict and, 138–40, 245n37
sewers, 87, 146, 148, 150, 189. *See also* Metrowater, Chennai; sanitation services
Shah, Mihir, 8–9
Singh, Manmohan, 244n17
Singh, Satyajit, 181–82
Siruvani Dam, 60, 138
Slum Clearance Board, Tamil Nadu (TNSCB), 153, 188–89
small towns, 88, 142, 178–82, 248n31. *See also* rural areas
Smart Cities Mission, 1, 2, 24, 87–88, 169
social movements, 99, 135–36, 230–31, 245n29. *See also* protests and protest movements
socioeconomic inequality. *See* inequality
Srisailam Reservoir, 129
state, 23, 25, 61, 73–81, 86, 195. *See also* bureaucratic agency; bureaucrats; state power
state capacity and effectiveness: institutional mechanisms for, 10; interstate water disputes and, 14–15, 117–18, 240n7 (intro.); studying dynamics of, 21; water markets as failure of, 173–74

state centralization. *See* centralization
state governments, 5–6, 9, 11–14, 18, 59, 89, 115, 119, 140, 179–80, 197, 230; interstate water disputes and, 99; JNNURM and, 87; Smart Cities Mission funding and, 88; social welfare impact of, 239n5; sovereignty, 119; staff reductions, 199; Supreme Court on financial distress of farmers and, 169–70; on urbanization of agricultural communities, 144–45; water resources and infrastructure authority of, 32–33; Water Users' Associations and, 69; World Bank and, 72, 86. *See also* central government; interstate water disputes; local governments
state power, 30, 63, 70–71, 89–95, 107, 128, 138–39, 161, 166, 174–78, 231; exercise of, through water reforms, global norms, national policies and, 81–89; governance of water and, 142, 226–27; large infrastructure projects as, 35, 87; liberalization policies, patronage, and, 200–201; of PWD's water bureaucracy in interstate negotiations, 53–54; in small towns and rural communities, 178–82; World Bank and, 81. *See also* centralization; corruption; patronage; political power
States Reorganisation Act (1956), 103–4
Sukthankar Committee, 242n6
Supreme Court, 3, 18, 59, 102, 107, 113–15, 131, 169–70; Inter-state River Water Disputes Act and, 244n16; interstate water disputes and, 13, 96, 99, 111; Mullaperiyar Dam conflict and, 131, 133–34, 135, 137; river-interlinking project and, 87, 94, 242n9
Swain, Ashok, 109
Swajaldhara Guidelines (2002), 84, 89

Tamil Nadu Centre for Public Interest Litigation, 170

Tamil Nadu Groundwater (Development and Management) Act (2003), 93–94, 160, 165
Tamil Nadu Labour Department, 204–5
Tamil Nadu Public Service Commission, 211
Tamil Nadu Public Works Department. *See* Public Works Department (PWD)
Tamil Nadu Public Works Department Association of Engineers, 249n14
Tamil Nadu Slum Clearance Board (TNSCB), 153, 188–89
Tamil Nadu Urban Development Fund (TNUDF), 91–92, 154, 178, 180, 248n30
Tamil Nadu Urban Finance and Infrastructure Development Corporation, 70
Tamil Nadu Urban Infrastructure Financial Services Limited (TNUIFSL), 24, 92, 178–79
Tamil Nadu Water Resources Consolidation Project, 76, 79, 89, 145–46, 147, 183. *See also* Water Resources Consolidation Projects (WRCPs)
Tamil Nadu Water Supply and Drainage Board (TWAD), 147, 148*fig*, 160, 162–64, 181, 248n31
Telangana, 100*map*, 129, 244n19
Telugu Ganga Project, 102, 121–25, 126*map*, 127–29, 131–32, 213, 244n24
Telugu speakers, 103–4, 121, 215
Tiwari, Piyush, 87
towns, small, 88, 142, 178–82, 248n31. *See also* rural areas
tribal groups, Narmada dam effects on, 75
tribunals, 96, 99, 111, 131. *See also* Cauvery Water Tribunal; Krishna Water Tribunal
tube well irrigation, 166

United Nations Development Programme (UNDP), 58, 67, 90, 157, 159
upward mobility, 65–66, 202, 208–9, 212

urban areas and urbanization, 3–5, 71, 83, 110, 148–49, 137, 169, 183, 187, 188*fig*., 206, 225, 234–37; competing and conflicting demands of rural areas and, 9, 67; interstate water disputes and, 99, 134; in Tamil Nadu, 17, 149*table*. *See also* Chennai; irrigation; metropolitan cities; poor people
urban governance, 6, 24–28, 141
Urban Infrastructure Development Scheme for Small and Medium Towns (UIDSSMT), 88
urban local bodies (ULBs), 83–85, 87–88, 91–92, 178–81, 229, 242n6
user fees, 176, 177, 180–81. *See also* cost recovery; revenue collection; water charges; water metering
Uttar Pradesh, 64–65, 89, 91

Vaddiraju, Anil Kumar, 6
Vaishnav, Milan, 18, 200
Veeranam Lake, 168
Vijayan, Pinarayi, 136
Village Water and Sanitation Committees, 182
violent protests, 106–7, 118. *See also* protests and protest movements

Wade, Robert, 64
Washbrook, David, 37–38
Washington Consensus, ix, 229
Water and Sanitation Program–South Asia (WSP-SA), World Bank, 77, 78
water body maintenance, 60, 87, 226
water charges, 48–49, 90. *See also* cost recovery; revenue collection; user fees; water metering
water disputes: colonial PWD's role and legacy in, 32, 50–51; politicization of, 115–16; postcolonial water governance and, 56–57. *See also* Cauvery (Kaveri) River dispute; interstate water disputes; Mullaperiyar Dam conflict

water governance, 4, 31–32, 49, 54–55, 85, 210, 231–33, 236–37; developmental state and, 55–63; postliberalization era and, 10–16; reforms and, 73–81, 239n4; World Bank and, 70. *See also* drinking water; groundwater; integrated water management; irrigation; politics of water

water mafias, 177

water markets, 15, 157, 159, 166–68, 171–72, 225; inequalities of class, caste, and gender and, 182–86

water metering, 176–77. *See also* cost recovery; revenue collection; user fees; water charges

water rates. *See* revenue collection

Water Regulatory Authority (WRA), 86

Water Resources Consolidation Projects (WRCPs), 77, 79, 90–91, 161. *See also* Chennai's water bureaucracy reforms; Tamil Nadu Water Resources Consolidation Project

Water Resources Organisation (WRO), 90, 147–48, 148*fig.*, 151–54, 162–63, 206

water scarcity: in Chennai, 58, 94, 132–42, 171, 176. *See also* droughts; famine

water services, 71, 74–77, 80, 176–77

water sharing, 53, 121–22

water tankers, 172–73, 173*fig.*, 207, 230

water tanks, 38–39, 41, 45–46, 49–50, 59–60

Water Users' Associations (WUAs), 69, 80, 83, 92–93 181, 242n14; gendered social norms and, 185–86; participatory irrigation management and, 88–89, 145–46

Weber, Max, 232

welfare of the people. *See* public interest; public welfare

wetlands development, 144, 151, 152–53, 246n6

women, 79, 184–86, 209, 249n15

World Bank, 3, 24–25, 69–70, 71–81, 86, 163–64, 15–76, 184–85, 229; Tamil Nadu and, 89–90; water sector funding by, 17, 145–46, 180

WRO. *See* Water Resources Organisation (WRO)

zamindars, 38, 50

www.ingramcontent.com/pod-product-compliance
Lightning Source LLC
Chambersburg PA
CBHW030610230426
43661CB00053B/1915